Catalogue of Watermarks in Italian Printed Maps, ca. 1540-1600

DAVID WOODWARD

CATALOGUE OF WATERMARKS IN ITALIAN PRINTED MAPS ca 1540-1600

The University of Chicago Press

David Woodward is the Arthur H. Robinson Professor of Geography at the University of Wisconsin, Madison. He is the editor of the *History of Cartography*, published by the University of Chicago Press

The University of Chicago Press, Chicago 60637
Casa Editrice Leo S. Olschki, Florence

Printed in Italy

05 04 03 02 01 00 99 98 97 96   1 2 3 4 5 6

ISBN 0-226-90727-9 (cloth)

A CIP record for this book is available from the Library of Congress.
This book is printed on acid-free paper.

CONTENTS

INTRODUCTION

Almost forty years ago, the dean of Italian bibliographers, Roberto Ridolfi, called for a systematic catalogue of watermarks in incunabula (RIDOLFI 1957, pp. 26-27). He realized that this was a huge and difficult task, requiring not only the photography of the watermarks in all the editions, but also of all the available examples of each edition. In making the case for such a reference work, he pointed to the difficulties of using Briquet's standard album (BRIQUET 1968). Although this contains more than 16,000 examples, its utility is seriously hampered by using hand tracing instead of photographic images; the subjectivity of this method (among other factors) has exposed the use of watermarks to severe criticism, most notably by Konrad Haebler (HAEBLER 1925). Despite the obvious care in copying, the subtleties of watermark images are usually such that it is usually impossible to know whether differences in the watermarks are due to different paper moulds or small irregularities in tracing. Consequently, efforts to match a given watermark with one in Briquet or other watermark manuals are usually met with frustration. Ridolfi, on the other hand, believed the minute variations in watermarks-visible only by some radiographic or photographic method-to be an advantage and not a drawback in identifying the source and chronology of paper used in book production. In the manual he envisioned, a complete set of photographic watermark images from fifteenth-century books would be an extraordinary tool in placing the many undated works into the chronology.

Unfortunately, Ridolfi's vision has not yet been fulfilled for incunabula, let alone the far more numerous books of the sixteenth century. Conor Fahy has remarked that for Italy, at the very least, local catalogues of watermarks occurring in printed books are needed.(FAHY 1995, p. 57) There have, indeed, been specialized studies that have employed radiographic or photographic methods and many more that have used the traditional tracing technique. The major sources of watermark images from the Italian Renaissance-BRIQUET 1968 [1907], ZONGHI 1953, BEANS 1938, and more recently BRUNO AND VIGNONI 1989, for example-have offered only hand tracings.

This catalogue is intended as a small contribution to the fund of objective images of sixteenth-century watermarks. For this reason, each entry includes the code specified in the International Standard for the Registration of Watermarks published by the International Association of Paper Historians, Provisional Edition, 1992. The catalogue's purpose is to provide researchers with radiographic and photographic reproductions of watermarks found in Italian printed maps (and a few prints) published in the approximate period 1540 to 1600. The images were gathered largely in 1979 and 1980 as essential evidence in reconstructing the Italian map trade of the sixteenth century while I was employed at The Newberry Library in Chicago. The core of the material was derived from a collection of Italian printed maps assembled by Franco Novacco and acquired by The Newberry Library, Chicago in 1967. It was supplemented by material lent to the library for the purpose of recording the watermarks, particularly a composite atlas and several separate printed maps in the stock of Kenneth Nebenzahl, Chicago, as well as several maps in the Roy V. Boswell Collection for the History of Cartography, California State University at Fullerton. The institutions that provided radiographs of watermarks in their collections are listed after the relevant entry and identified in the list of abbreviations.

The 335 images reproduced here have been selected from more than 1100 images collected over the course of study. On average, therefore, about three or four examples of a particular paper mould were originally gathered. In some cases, only one example was found; in others, many more maps printed on paper from a single paper mould have been collected. Mould number 92, for example, is represented by no fewer than twenty-five impressions.

Due to various working conditions, different thicknesses of paper, and different methods of recording the watermarks, the quality of the images has not always been optimal. The clearest examples have been selected for reproduction, and those of lower quality have been included only if they allow identification of the paper mould.

In any work involving watermarks, it is necessary to understand that it is the paper mould, not the paper, and certainly not the map or print, that is being identified by the watermark. Curt Bühler pointed out succinctly that any study that purports to use watermarks to date early documents must seriously consider five hypothetical assumptions:

(1) a mould for making paper had a limited life-ranging from a minimum of four months to a maximum span of four years.

(2) the mould was used continuously-there were no extensive gaps of time in its use.

(3) all the paper produced by a mould was used up in two or three years-it was not used by a printer over a long span of years.

(4) the paper was used by the printer in the same order in which it came from the mould.

(5) stocks of paper were not accumulated-it was too expensive and thus required fast turnover. (BÜHLER 1973, pp. 2-3)

If one is to accept these assumptions in an absolute sense, they are of course difficult to dismiss. Some are certainly more compelling than others. Thus, while it is difficult to imagine any scenario in which a mould was *always* used continuously, stocks *never* accumulated, or paper was *always* used in the same order from ream, it appears from the careful studies of Allan Stevenson and others that compelling evidence exists for the short life of paper moulds. Stevenson showed that paper moulds go through identifiable stages in their lives analogous to the states of a printing plate (STEVENSON 1967). The mark may become increasingly distorted with use as the mould is jostled or as excess pulp is brushed from it. More dramatically, if the mark is situated between chain lines and not sewn to a chain line passing through it, the sewing wires tend to become loose with age and the mark moves slowly to the left in relation to the chain lines. Stevenson even estimated the rate of movement as averaging about a millimeter a month, a distance certainly discernible on a radiograph (STEVENSON 1967, p. 251). At this rate, the life of a watermark before it became severely distorted could be quite short.

The dates listed for the maps in this catalogue are the dates engraved on the plates from which the maps were printed. They are not the dates of the impressions. Thus, a map from a plate dated 1546, for example, might be printed on paper many years later. These printed dates, therefore, represent a *terminus a quo* for each paper mould, but not a *terminus ad quem*. For the latter, it is necessary to search for other evidence. This might include internal bibliographical evidence provided by the atlas in which the maps were bound, the presence of offset images on the verso, or a change in the state of the plate. For example, it has been shown that the two states of a map of North America by Paolo Forlani (1565 and 1566) bear characteristically different watermarks (WOODWARD 1994), thus establishing the *terminus ad quem* of watermark number 128 as 1565.

What if a given batch of paper were left standing in a distributor's warehouse or printer's stock room long after manufacture? Stevenson countered this objection by pointing out that, with paper making up half the printer's cost of materials, it would have been uneconomic to have such capital inactive for a long period (STEVENSON 1967, p. 49). In the context of maps and prints, however, this evidence is less compelling, for while a ream or two might be sufficient for the total number of impressions from a copper plate, maybe 20 or 25 reams would have been necessary for the average book.

Despite the pitfalls of watermark evidence, I agree with Ridolfi that the skepticism toward the dating of paper using the evidence of paper moulds is perhaps not entirely warranted. The watermark data gathered seems to confirm that it is unusual indeed to find a given watermark associated with a map or print substantially out of the period in which the mould was probably in use. For example, the siren watermark pair numbers 91 and 92 contain maps dated 1557, 1559, and every year from 1562 to 1570. The earlier maps are from plates that were still available in the late 1570s, and as I have argued elsewhere (WOODWARD 1985, 1987) it is very likely that the life of this pair of watermarks can be narrowed to 1568-1570 or even 1569-1570. As more and more examples are found (assuming of course that no map is dated after 1570), the likelihood of fixing the *terminus ad quem* at 1570 is increased.

It has also been shown-using a combination of radiography and element analysis in paper-that there is a high correlation with the use of a given watermark and a particular batches of paper (WOODWARD 1990b). It is unlikely that different papermakers using different recipes of pulp employed the same paper mould. In other words, it is unusual indeed for a watermark to appear grossly out of context, as would happen if paper moulds (with their watermarks intact) would be transferred from one papermaker to another.

While the evidence of paper is useful, therefore, it needs to be used in combination with other bibliographical evidence, such as plate histories and offsets, in order to be most effective. WOODWARD 1991 provides an example of how watermarks can be used with offset evidence to reconstruct families of plates that were printed in the same place at the same time.

Needham has summarized the main methods of watermark reproduction available (NEEDHAM 1988). Of the three main methods available for this catalogue-ultraviolet Dylux (GRAVELL AND MILLER 1979), the Ilkley method using high-speed duplicating film (SCHOONOVER 1987), and beta radiography-the last was found to be the most effective. Its advantage lay in its creation of a permanent negative image from which prints could be taken, coupled with the advantage that the detail on the map or print is not reproduced. For the maps in the Doria Atlas, the Ilkley method was used. Other methods have of course been proposed and some have been employed on a large scale. Examples include an X-ray method in use in Amsterdam

(LAURENTIUS ET AL 1992). A new method-the Polaroid method-bears even more promise in its combination of cheap portable equipment, safety, clarity, and flexibility, but this method was not available during the period of this project.

Beta radiography has a fairly long history in bibliographical analysis (SIMMONS 1961, STEVENSON 1967, TANSELLE 1971, pp. 48-50, SPECTOR 1987, NEEDHAM 1988). The large majority of the examples in this catalogue were obtained at the Newberry Library with a carbon 14 impregnated plastic sheet (16 x 23 cm) emitting beta rays at the rate of approximately 50 millirems per hour (a dental X-ray delivers 20 millirems on one tenth of a second). The plate is placed on one side of the map to cover the watermarked area. A piece of X-ray film was placed on the other side. Exposure time was between 10 and 15 minutes depending on the thickness of the paper. Since the paper is less dense in the image area of the watermark (where the wires of the mould have come in contact with the pulp), the beta rays pass more easily through these areas and consequently expose the X-ray film more intensively. The result is a negative on which the watermark image appears dark. Routine photographic contact prints (such as those used here) can then be reproduced on which the watermark image appears light (WOODWARD 1979).

The main identification tool to the maps in which watermarks were found is TOOLEY 1939. As is well known, this list is by no means complete and is in dire need of updating and correcting, but it is the most useful general list that exists. A manuscript of a revised list by Tooley containing half as many again additions was in preparation, but this disappeared after his death and may have been destroyed. For those maps not listed in Tooley, more information and bibliographical references have been provided to enable identification. In addition, the collections in which the items are found and their call numbers have been provided.

A work of this type, undertaken intermittently over a long period and involving unusual access to original material with complex equipment, owes its existence to the help of many individuals and institutions. Without the cooperation of the Newberry Library, where I was employed between 1969 and 1980, the work would have been impossible. It is a pleasure to acknowledge the help of Bill Towner, Jim Wells, Paul Banks, John Tedeschi, John Aubrey, Bob Karrow, David Buisseret, Jim Akerman, and Paul Gehl at various stages of the work. Special thanks are due to Maria Holden, who did the bulk of the beta radiography in 1979 and 1980, and to Ken Cain, who completed the radiography of the Newberry's sixteenth-century printed maps more recently. Likewise, I must thank Kenneth Nebenzahl, Roy V. Boswell, and Arthur Holzheimer for their generosity in lending their material for radiography.

Research support for this project has been particularly welcome. A John Simon Guggenheim Fellowship in 1977-78 allowed me to start work on the context of the sixteenth-century Italian map trade, for which the watermark evidence has been a necessary tool. A fellowship at the Hermon Dunlap Smith Center for the History of Cartography at The Newberry Library in 1989-90 and a British Academy Fellowship in London in 1993 allowed further opportunities to work intermittently on the list. Support from the Graduate School of the University of Wisconsin-Madison for student hourly help, supplies, and expenses in between 1983 and 1988 is also gratefully acknowledged. Students who helped with the project at various stages included-in chronological order-Mark Olsen, Matthew Edney, Josh Hane, and Michael Paré. Several colleagues in various fields smoothed the way or provided helpful discussions, including Cristina Armeni, Luigi Balsamo, Tony Campbell, Ed Dahl, Conor Fahy, Albert Ganado, Paul Gehl, Paul Needham, David Park, David Schoonover, and G. Thomas Tanselle. Several cooperating institutions must also be gratefully acknowledged, such as the Finnish Pulp and Paper Institute, the New York Public Library, Helsinki University Library, the National Map Collection of the National Archives of Canada, and Sotheby Parke Bernet. As always, my family has been most understanding in my pursuit of a specialized technique that must sometimes have appeared bizarre.

EXPLANATION OF ENTRIES

The entries consist of:

A) A GENERAL DESCRIPTION OF THE WATERMARK MOULD, INCLUDING

1. The watermark mould number.
2. A descriptive title, organized in the order recommended by the International Standard for the Registration of Watermarks published by the International Association of Paper Historians, Provisional Edition, 1992.
3. The size of the salient feature, such as "circle 52 mm."
4. A notation of special characteristics.
5. A code such as "A1{p05}-U1" following the specifications in the International Standard for the Registration of Watermarks published by the International Association of Paper Historians, Provisional Edition, 1992.

B) EACH SUBENTRY REFERS TO A SPECIFIC IMPRESSION OF A MAP OR PRINT, WITH THE FOLLOWING INFORMATION:

1. Date printed on impression; n.d. signifies that no date is printed.
2. The title or area of the map [in square brackets when summarized in English].
3. Place of publication printed on impression; n.pl. signifies that no place is printed on the map; [place in square brackets is estimated from other evidence].
4. Name of printer or publisher appearing on impression; n.pr. signifies that no publisher is printed on the map; [printer in square brackets is estimated from other evidence].
5. Map number from Tooley's list of Italian printed maps (see bibliography).
6. Other references, when not listed in Tooley or to help to identify the impression.
7. The institution and call number where the impression is found.

An entire entry in italics indicates the watermark image chosen for reproduction in this publication.

ABBREVIATIONS

Boswell	=	Composite atlas of provenance George H. Beans – Henry Stevens – Roy V. Boswell – Kenneth Nebenzahl
CSUF	=	Roy V. Boswell Collection for the History of Cartography, California State University, Fullerton
Doria Atlas	=	Composite atlas known as the *Doria Atlas*, described in Sotheby Catalogues for 15 April 1980 and 27 September 1988
GHB	=	Separate maps belonging to George H. Beans acquired by Roy V. Boswell
ISRW	=	International Standard for the Registration of Watermarks
KN	=	Kenneth Nebenzahl, Glencoe, Illinois
(ms)	=	watermark in marginal strip
NC	=	Nordenskiöld Collection, Helsinki University Library. Radiographs supplied by the Finnish Pulp and Paper Institute, Helsinki. See MICKWITZ 1979-84
n.d.	=	undated
n.i. TOOLEY	=	not found in TOOLEY 1939
NL	=	The Newberry Library, Chicago. Call numbers beginning with 2F, 4F, 6F are in the Franco Novacco Collection
NMC	=	National Map Collection (National Archives of Canada)
n.pl.	=	no place of publication indicated
n.pr.	=	no printer indicated
NYPL	=	New York Public Library, New York
(sim.)	=	similar to
VCA	=	Venetian composite atlas in The Newberry Library, call number Ayer *135 L2 1575 AI

The orientation of the watermarks does not always read as one would read the watermark on the original paper mould. Information about the «mould side» of the paper was not gathered at the time the radiographs were made. Readers should be aware that the watermark they are matching may be in mirror image to the one published here. In cases where the orientation is obvious, as when letters or numbers are included, the image is correctly oriented.

REFERENCES TO INTRODUCTION AND CATALOGUE

BEANS 1938 = GEORGE H. BEANS, *Some Sixteenth Century Watermarks Found in Maps Prevalent in the "IATO" Atlases.* Edited by GEORGE H. BEANS. «Publications of the George H. Beans Library» XIII, Jenkintown, The George H. Beans Library, 1938.

BORRONI SALVADORI 1980 = FABIA BORRONI SALVADORI, *Carte, piante e stampe storiche delle raccolte lafreriane della Biblioteca Nazionale di Firenze.* «Indici e Cataloghi», Nuova serie XI, Rome, Istituto Poligrafico e Zecca dello Stato, 1980.

BRIQUET 1968 [1907] = CHARLES M. BRIQUET, *Les filigranes. Dictionnaire historique des marques du papier dès leur apparition vers 1282 jusqu'en 1600. A facsimile of the 1907 edition with supplementary material contributed by a number of scholars,* Edited by Allan Stevenson. Amsterdam-Hilversum, Paper Publications Society, 1968.

BRUNO AND VIGNONI 1989 = GIGLIO BRUNO AND ILO VIGNONO, *Incunaboli, cinquecentine e loro filigrane, impronte, P.D.P [parametro di paginazione], P.A.G.I. [parametro d'arte grafica e incisoria] della Biblioteca Capitolare d'Ivrea*, Ivrea, Bolognino Editore, 1989. 2 vol. Watermarks reproduced Vol. 2, pp. 1225-1330.

BÜHLER 1973 = CURT F. BÜHLER, *Last Words on Watermarks*, «Papers of the Bibliographical Society of America» LXVII, 1973, pp. 116.

FAHY 1995 = CONOR FAHY, *Roberto Ridolfi e lo studio bibliologico della carta*, «La Bibliofilia» XCVII, 1995, pp. 35-57.

GANADO 1993 = ALBERT GANADO, *The Forlani-Zenoi Town Books of 1567: A Description and Analysis*, In PETER H. KOHL AND PETER H. MEURER, eds., «Florilegium Cartographicum. Beiträge zur Kartographiegeschichte und Vedutenkunde des 16. bis 19. Jahrhunderts Fritz Hellwig zu Ehren», Leipzig, Verlag Dietrich Pfaehler, 1993, pp. 21-34.

GRAVELL AND MILLER 1979 = THOMAS L. GRAVELL AND GEORGE MILLER, *A Catalogue of American Watermarks 1690-1835*, New York, Garland Publishing, 1979.

HAEBLER 1925 = KONRAD HAEBLER, *Handbuch der Inkunabelkunde,* Leipzig, K. W. Hiersemann, 1925.

HEAWOOD 1924 = EDWARD HEAWOOD, *The Use of Watermarks in Dating Old Maps and Documents*, «Geographical Journal» LXIII, 1924, pp. 391-412 and appendix of two plates.

HEAWOOD 1929 = EDWARD HEAWOOD, *An Undescribed Lafreri Atlas and Contemporary Venetian Collections*, «Geographical Journal» LXXII, 1929, pp. 359-69.

HEAWOOD CAT = EDWARD HEAWOOD. *Watermarks, Mainly of the 17th and 18th Centuries,* Hilversum, Paper Publications Society, 1950.

HEYDEN 1993 = H. A. M. VAN DER HEYDEN, *Matteo Florimi (-1613): Landkarten- und Stadtplanverleger in Siena*, In PETER H. KOHL AND PETER H. MEURER, eds., «Florilegium Cartographicum. Beiträge zur Kartographiegeschichte und Vedutenkunde des 16. bis 19. Jahrhunderts Fritz Hellwig zu Ehren», Leipzig, Verlag Dietrich Pfaehler, 1993, pp. 117-30.

HÜLSEN 1921 = CHRISTIAN HÜLSEN, *Das 'Speculum Romanae Magnificentiae des Antonio Lafreri'*, In «Collectanea variae doctrinae Leoni S. Olschki», Munich, Rosenthal, 1921, pp. 121-70.

LAURENTIUS ET AL 1992 = THEO LAURENTIUS, HARRY M. M. VAN HUGTEN, ERIK HINTERDING AND JAN PIET FILEDT KOK, *Het Amsterdamse onderzoek naar Rembrandtspapier: radiografie van de watermerken in de etsen van Rembrandt*, «Bulletin van het Rijksmuseum» XL, 1992, pp. 353-84.

MASSARI 1987 = STEFANIA MASSARI AND FRANCESCO NEGRI ARNOLDI, *Arte e scienza dell'incisione da Maso Finiguerra a Picasso*, Rome, 1987.

MICKWITZ 1979-84 = ANN-MARI MICKWITZ AND LEENA MIEKKAVAARA, *The A. E. Nordenskiöld Collection in the Helsinki University Library: Annotated Catalogue of Maps made up to 1800,* Stockholm, Almqvist & Wiksell International, 1979-84.

MOSIN 1973 = VLADIMIR MOSIN, *Anchor Watermarks,* «Monumenta Chartae Papyraceae Historiam Illustrantia or Collection of Works and Documents Illustrating the History of Paper» XIII, ed. J. S. G. Simmons, Amsterdam, The Paper Publications Society (Labarre Foundation), 1973.

MUSEO COMUNALE DI GUBBIO 1993 = CRISTINA ARMENI AND CRISTINA FALCUCCI, *Museo Comunale di Gubbio: Incisioni*, Perugia, Electa Editori Umbri, 1993.

NEEDHAM 1988 = PAUL NEEDHAM, *The Study of Paper from an Archival Point of View*, «IPH [International Association of Paper Historians] Yearbook» VII, 1988, pp. 122-36.

PICCARD 1961 = G. PICCARD, *Kronenwasserzeichen. Findbuch I der Wasserzeichenkartei Piccard in Hauptstaatsarchiv Stuttgart,* Stuttgart, 1961.

RIDOLFI 1957 = ROBERTO RIDOLFI, *Le filigrane dei paleotipi: Saggio metodologico*, Florence, Leo S. Olschki, 1957.

ROBERTS 1988 = JANE ROBERTS, *A Dictionary of Michelangelo's Watermarks,* Milan, Olivetti, 1988.

SCACCIA SCARAFONI 1939 = CAMILLO SCACCIA SCARAFONI, *Le piante di Roma possedute dalla Biblioteca dell'Istituto e dalle altre biblioteche governative della città,* Rome, Istituto di archeologia e storia dell'arte, 1939.

SCHOONOVER 1987 = DAVID SCHOONOVER, *Techniques of Reproducing Watermarks: A Practical Introduction*, In «Essays In Paper Analysis», ed. Stephen Spector, Washington, D.C., Associated University Presses, 1987, pp. 154-67.

SCHULZ 1970 = JUERGEN SCHULZ, *The Printed Plans and Panoramic Views of Venice (1486-1797)*, «Saggi e memorie di storia dell'arte (Fondazione Giorgio Cini)» VII, Florence, Leo S. Olschki, 1970.

SIMMONS 1961 = J. S. G. SIMMONS, *The Leningrad Method of Watermark Reproduction*, «The Book Collector» X, 1961, pp. 329-30 and plates.

SPECTOR 1987 = STEPHEN SPECTOR, ed., *Essays In Paper Analysis*, Washington, D.C., Associated University Presses, 1987.

STEVENSON 1961 = ALLAN STEVENSON, *Observations on Paper as Evidence*. «University of Kansas Library Series» XI, Lawrence, University of Kansas, 1961.

STEVENSON 1967 = ALLAN STEVENSON, *The Problem of the Missale Speciale*, London, The Bibliographical Society, 1967.

TANSELLE 1971 = G. THOMAS TANSELLE, *The Bibliographical Description of Paper*, «Studies in Bibliography» XXIV, 1971, pp. 27-67, esp. pp. 48-50

TOOLEY 1939 = RONALD VERE TOOLEY, *Maps in Italian Atlases of the Sixteenth Century, Being a Comparative List of the Italian Maps issued by Lafreri, Forlani, Duchetti, Bertelli and Others, Found in Atlases*. «Imago Mundi» III, 1939, pp. 12-47.

WOODWARD 1979 = DAVID WOODWARD, *Watermark Radiography at the Newberry Library*, «Mapline» XV, 1979, pp. 1-2.

WOODWARD 1985 = DAVID WOODWARD, *New Tools for the Study of Watermarks on Sixteenth-Century Italian Printed Maps: Beta Radiography and Scanning Densitometry*, In «Imago et Mensura Mundi: Atti del IX Congresso Internazionale di Storia della Cartografia» II, ed. Carla Clivio Marzoli, Rome, Istituto della Enciclopedia Italiana, 1985, pp. 541-52.

WOODWARD 1987 = DAVID WOODWARD, *The Analysis of Paper and Ink in Early Maps: Opportunities and Realities*, in STEPHEN SPECTOR, ed., «Essays in Paper Analysis», Washington, D.C., Associated University Presses, 1987, pp. 200-21.

WOODWARD 1990a = DAVID WOODWARD, *The Maps and Prints of Paolo Forlani: A Descriptive Bibliography*, «Hermon Dunlap Smith Center for the History of Cartography Special Publication» IV, Chicago, The Newberry Library, 1990.

WOODWARD 1990b = DAVID WOODWARD, *The Correlation of Watermark and Paper Chemistry in Sixteenth Century Italian Printed Maps*, «Imago Mundi» XLII, 1990, pp. 84-93.

WOODWARD 1991 = DAVID WOODWARD, *The Evidence of Offsets in Renaissance Italian Maps and Prints*, «Print Quarterly» VIII, 1991, pp. 235-51.

WOODWARD 1994 = DAVID WOODWARD, *The Forlani Map of North America*, «Imago Mundi» XLVI, 1994, pp. 29-40.

ZONGHI 1953 = AURELIO ZONGHI, AUGUSTO ZONGHI, AND A. F. GASPARINETTI *Zonghi's Watermarks*, «Monumenta chartae papyraceae historiam illustrantia» III, Hilversum, The Paper Publications Society, 1953.

CATALOGUE

1. Hands clasped in circle; circle 49 x 50 mm. A1{p05}-U1

1590 ca; [Hispaniola]; Venice, Paolo Forlani (Petri de Nobilibus); TOOLEY 85, Boswell 222r.

2. Pilgrim, vertical staff, in circle A; circle 51 x 50 mm; central wire bent. A4/1[R1/1]-U1

1552; [Bird's eye view of Rome]; Rome, Michele Tramezzino; TOOLEY 484, Doria Atlas 70.

1565; Iodoco á Meggen Lucernati Praetorianorum Praefecto Antonio Salamanca; Rome, Antonio Lafreri (Jacob Bos); n.i. TOOLEY, NC 2:130:32 right.

1565; [Malta]; Rome, Antonio Lafreri; TOOLEY 360, NL 2F43.

***1566**; [Naples]; Rome, Antonio Lafreri; TOOLEY 408, NL 4F232.*

1566; [France]; Venice, Bolognino Zaltieri; TOOLEY 211, Doria Atlas 23.

1567; [Brabant]; Venice, Bolognino Zaltieri; TOOLEY 145, NL 4F152.

1567; [Holland]; [Venice], Bolognino Zaltieri; TOOLEY 302, GHB 212.

1568; [Poland, 2 sheets]; Venice, Paolo Forlani; TOOLEY 454-55, Doria Atlas 38.

1570; [Territory of Cremona]; Venice, Paolo Forlani; n.i. TOOLEY, Doria Atlas 62.

1570; [Egypt]; n.pl., n.pr.; TOOLEY 194, Doria Atlas 98.

n.d. Regno di Napoli; n.pl.,n.pr.; TOOLEY 404, NC 2:130:41.

n.d. [Zealand]; n.pl., n.pr.; TOOLEY 602, NL 2F212.

n.d. [Jerusalem]; n.pl., Stefano Dupérac?; TOOLEY 341, NL 2F232r.

n.d. [Jerusalem]; n.pl., Stefano Dupérac?; TOOLEY 341, NL 2F232v.

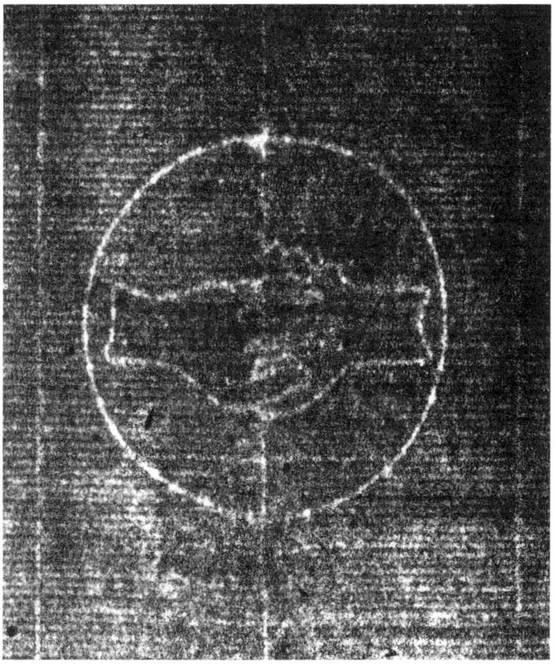

1

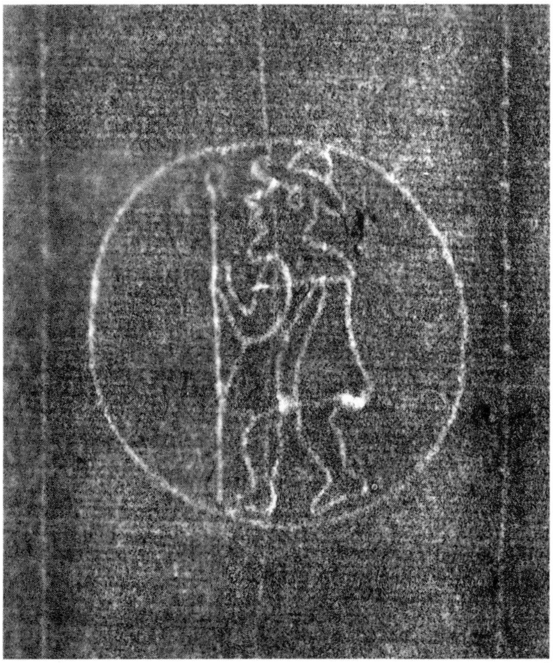

2

3

4

3. Pilgrim, vertical staff, in circle B; circle 52 x 55 mm. A4/1[R1/1]-U1

***1572* ca**; *Geografia. Tavole Moderne di Geografia; Rome, Antonio Lafreri; Tooley [unnumbered first page],* NC *2:130: title page.*

1561; [Italy]; Rome, Antonio Lafreri (Giacomo Gastaldi); Tooley 328, NC 2:130:33 left.

4. Pilgrim, vertical crook, in circle or oval A; circle 52 mm; hand on crook at split. A4/1[R1/1]-U1 Other manuals: Beans 38 (sim.)

1565; *[Malta]; Rome, Antonio Lafreri; Tooley 377, NL 2F44.*

1567; [Messina]; Rome, Antonio Lafreri; Tooley 388, NL 4F227.

n.d. [Taprobana (Sri Lanka)]; n.pl., n.pr.; Tooley 547, NL 2F235.

n.d. [Algiers]; n.pl., n.pr.; Tooley 97, NL 2F239.

n.d. [North America – New France]; [Venice, Paolo Forlani]; Tooley 81, NL 4F56 (bottom missing).

n.d. [Guines]; n.pl., n.pr.; Tooley 294, NL 4F100 (top half).

5. Pilgrim, vertical crook, in circle or oval B; circle 53 x 54 mm; hand on crook below split; arm of crook curved. A4/1[R1/1]-U1

1565; [Malta]; Rome, Antonio Lafreri; TOOLEY 360, GHB 233.

1565; [Malta]; Rome, Antonio Lafreri; TOOLEY 374, NL 2F38.

***1565**; [Malta]; Rome, Antonio Lafreri; TOOLEY 377, GHB 243.*

1565 ca; [Cuba 26 x 19 cm]; n.pl., n.pr.; n.i. TOOLEY, NL 2F14.

1567; [Boulogne]; n.pl., n.pr.; TOOLEY 140, CSUF ac 1324.

1567; [Calais]; n.pl., n.pr.; TOOLEY 152, CSUF 1319.

1590 ca; [Pozzuoli eruption 29 Sept. 1538 (Campania)]; n.pl., Petri de Nobilibus; TOOLEY 458, NL 4F344.

n.d. [Guines]; n.pl., n.pr.; TOOLEY 294, NL 4F100 (left third).

6. Pilgrim, vertical crook, in circle or oval C; circle 46 x 47 mm; hand on crook below split; arm of crook curved. A4/1[R1/1]-U1

***1558**; [Belgium]; Rome, Michele Tramezzino (Pirro Ligorio, Sebastiano di Re); TOOLEY 132, NL 4F146.*

5

6

7

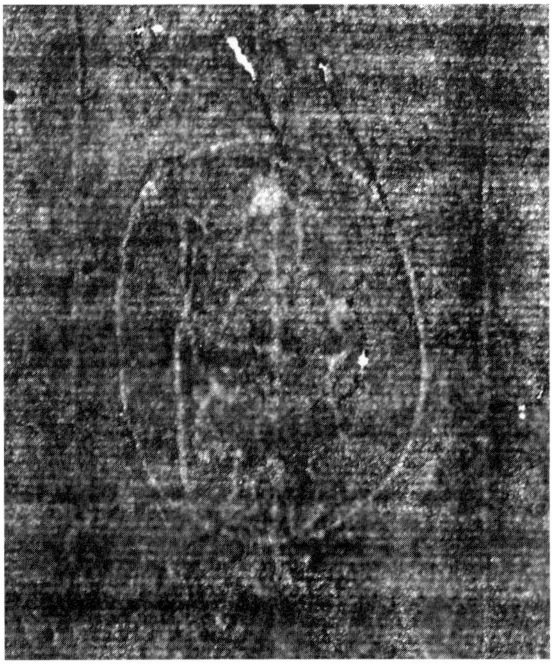

8

7. Pilgrim, vertical crook, in circle or oval D; circle 51 x 55 mm, bent; hand on crook above split; arm of crook bent back almost to staff; very bent wires. A4/1[R1/1]-U1

1563; [Gelderland]; Rome, Antonio Lafreri (Paolo Forlani); Tooley 234, NC 2:130:18 left margin.

1564; [Corfu]; Venice, Ferrando Bertelli; Tooley 165, Doria Atlas 93.

1566; *[Palestine]; Venice, Paolo Forlani; Tooley 436, NL 2F226.*

1566; [Holy Land]; Venice, Paolo Forlani; Tooley 436, Doria Atlas 106.

1570; [Asia Minor – Natolia and Caramania]; Venice, Bolognino Zaltieri; Tooley 66, Doria Atlas 104.

n.d. [Flanders]; Rome, Antonio Lafreri (Claudio Duchetti); Tooley 205, NC 2:130:16 left margin.

n.d. [Bohemia, 2 sheets]; Venice, Bolognino Zaltieri; Tooley 138, Doria Atlas 47b.

n.d. [Kingdom of Naples]; n.pl., n.pr.; Tooley 404, GHB 244/2.

n.d. [Messina]; n.pl./printer; Tooley 388, NL 4F227l (ms).

8. Pilgrim, vertical crook, in circle or oval E; oval 42 x 53 mm; crude. A4/1 [R1/1]-U1

n.d. *[Ancona]; n.pl., n.pr.; Tooley 108, NL 2F114.*

9. Pilgrim, vertical crook, in shield 70 mm tall. A4/1[R1/1]-M13

1545 ca; [Rome (Piazza S. Pietro)]; Rome, Antonio Lafreri; n.i. TOOLEY, NL 4F260.

1570; *[Grand Duke of Tuscany's visit to Rome, 18 February 1570]; n.pl., n.pr.; n.i. TOOLEY, NL 2F138.*

10. Pilgrim, crook over shoulder, in circle or oval A; circle 48 x 50 mm. A4/1[R1/1]-U1. Other manuals: BEANS 62 (sim.)

1558; *[Sicily]; Rome, Vincenzo Lucchino; TOOLEY 516, Boswell 213r.*

9

10

11

12

11. Pilgrim, crook over shoulder, in circle or oval B; circle 41 x 42 mm**;** crook almost horizontal; A4/1[R1/1]-U1

***1585* ca**; [*Augsburg*]; *n.pl.*, [*Claudio Duchetti*]; *Tooley 119, Boswell 239r.*

12. Pilgrim, crook over shoulder, in circle or oval C; circle 42 x 43 mm; crook touching circle. A4/1[R1/1]-U1

1569; [*Orvieto*]; *n.pl., n.pr.; Tooley 424, Boswell 240r.*

1575?; [La Rochelle]; n.pl., n.pr. (Antonio Lafreri's name erased from earlier state); Tooley 469 (state 2), Boswell 243r.

13. Pilgrim, crook over shoulder, in circle or oval D; circle 46 mm. A4/1[R1/1]-U1

1595; *Asia nova descriptio; Venice, Andrea Bertelli; n.i.* TOOLEY, *NL 2F217.*

14. Pilgrim, crook over shoulder, in circle or oval E; circle 45 x 46 mm. A4/1[R1/1]-U1

1585 ca; [*Civitella*]; *n.pl., n.pr. (Claudio Duchetti);* TOOLEY *154, Boswell 250r.*

13

14

15

16

15. Pilgrim, crook over shoulder, in circle or oval F; circle 50 x 52 mm. A4/1[R1/1]-U1

***1592* ca**; [*Mirandola*]; *n.pl., n.pr.; Tooley 394 (without A S imprint), Boswell 251r.*

16. Pilgrim, crook over shoulder, in circle or oval G; oval 46 x 50 mm, bent; crude. A4/1[R1/1]-U1

1575; [*Rome*]; *Rome, Antonio Lafreri; n.i. Tooley, NL 2F139.*

17. Pilgrim, crook over shoulder, in circle, under six-pointed star A; circle 44 x 45 mm; total height 65 mm. A4/1 [R1/1]-U1[J5/4]

__1585__; Algeri; Rome, Claudio Duchetti (Ambrogio Brambilla); n.i. T*OOLEY*, *CSUF ac 1323.*

18. Pilgrim, crook over shoulder, in circle, under six-pointed star B; circle 43 x 46 mm; total height 69 mm; crude. A4/1[R1/1]-U1[J5/4]

__1589__; [*British Isles*]; *Rome, Marcus Clodianus;* T*OOLEY 274, Boswell 205r.*

17

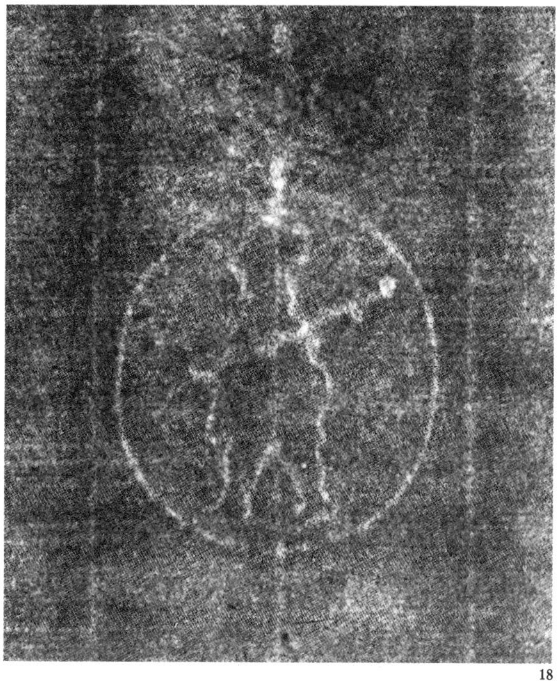

18

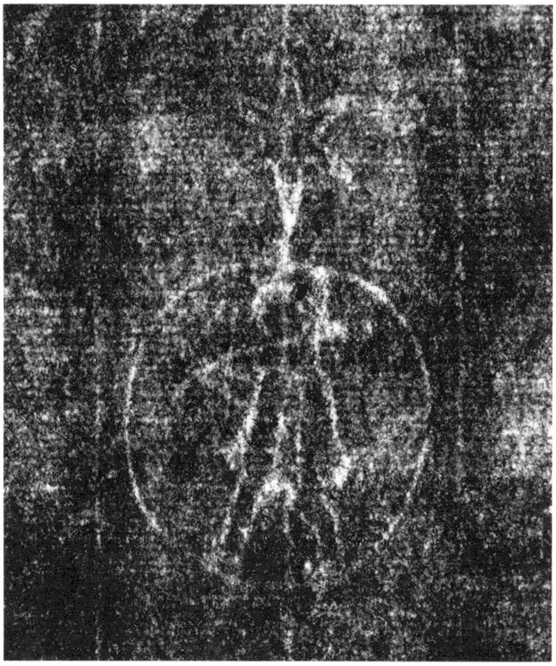

19

20

19. Pilgrim, crook over shoulder, in circle, under six-pointed star C; circle 41 x 43 mm; total height 70 mm. A4/1 [R1/1]-U1[J5/4]

n.d. *[Tuscany]; n.pl., n.pr.; Tooley 567, GHB 859.*

20. Pilgrim, crook over shoulder, in circle, under six-pointed star D; circle 48 x 49 mm; total height 69 mm. A4/1 [R1/1]-U1[J5/4]. Other manuals: Beans 74

1570; [Cyprus]; n.pl., Claudio Duchetti; Tooley 186, Boswell 219r.

1570; [Crete]; n.pl., Claudio Duchetti; Tooley 178, Boswell 220r.

1570; [Rhodes]; n.pl., Claudio Duchetti; Tooley 465, Boswell 221r.

1570; [Majorca]; n.pl., Claudio Duchetti; Tooley 357, Boswell 208r.

***1585** ca; [Zealand, Denmark]; n.pl., n.pr.; Tooley 602, Boswell 203r.*

1585 ca; [Ireland]; Rome, Claudio Duchetti; Tooley 319, Boswell 204r.

1585 ca; [Corfu]; n.pl., n.pr.; Tooley 163, Boswell 217r.

21. Pilgrim, crook over shoulder, in circle, under six-pointed star E; circle 48 x 51 mm; total height 81 mm. A4/1 [R1/1]-U1[J5/4]. Other manuals: BEANS 41

1570; [Minorca]; n.pl., Claudio Duchetti; TOOLEY 393, Boswell 209r.

1585 ca; [Elba]; n.pl., n.pr.; TOOLEY 196, Boswell 211r.

1585 ca; *[Sardinia]; n.pl., n.pr.; TOOLEY 512, Boswell 212r.*

22. Pilgrim, crook over shoulder, in circle, under six-pointed star F; circle 42 mm ; total height 64 mm. A4/1[R1/1]-U1[J5/4]

1561 ca; *[Greece]; n.pl., Sebastianus Clodiensis (Franciscus Salamanca); TOOLEY 284, NL 4F358.*

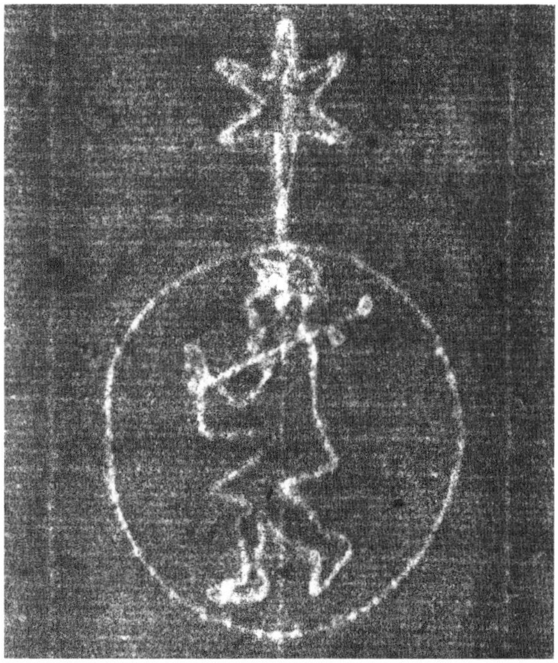

21

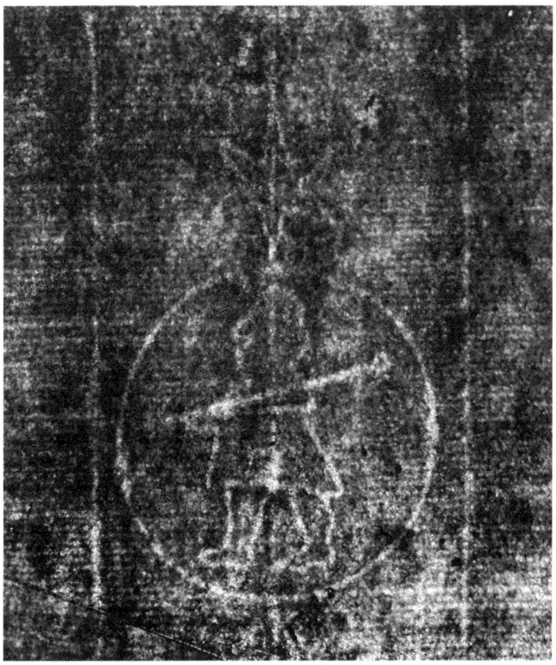

22

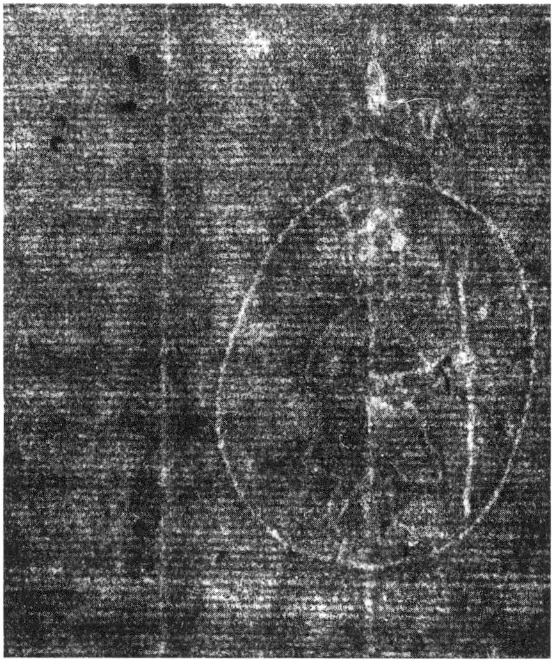

23

23. Pilgrim with vertical crook, in circle under crown with concave base; circle 43 x 51 mm; total height 78 mm. A4/1[R1/1]-U1 Other manuals: BEANS 64 (sim.)

***1590* ca**; *[Siena]; n.pl., Matteo Florimi; n.i. TOOLEY, VAN DER HEYDEN 43 OR 44, NL 4F278.*

24

24. Kneeling man and cross (large) in broad shield (74 mm tall). A4/1 S1[M13]. Other manuals: BORRONI SALVADORI 72 (sim.)

1558; *[Gelderland]; n.pl., Michele Tramezzino (Jacob Bos); TOOLEY 232, Boswell 183r.*

25. Kneeling man holding a small cross in narrow shield (68 mm). A4/1 S1[M13]

1618; [*Macerata (dedicated to Cardinal di Ascoli)*]; *n.pl., Floriani fecit.; n.i.* TOOLEY, *NL 4F224.*

26. Kneeling man holding a small cross in shield (37 x 70 mm). A4/1-S1[M13]

n.d. [*Brescia*]; *Giovanni Orlandi, Henricus van Schoel; n.i.* TOOLEY, *NL 4F216.*

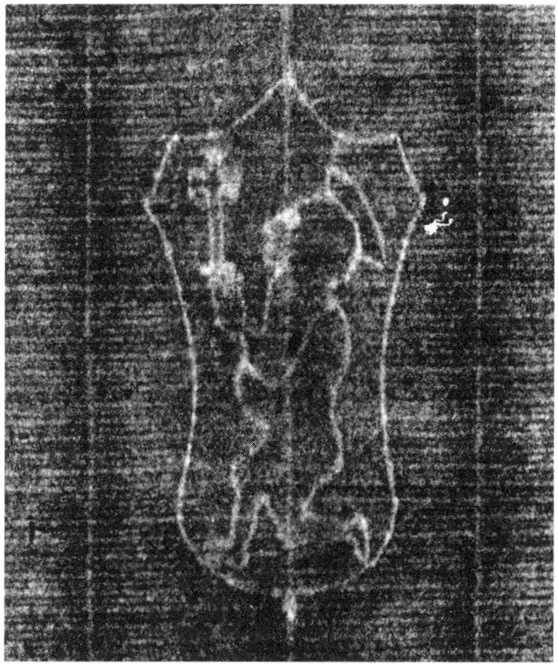

25

26

27

28

27. Kneeling man and cross in shield (40 x 70 mm). A4/1-S1[M13]

1591; *[Naples]; n.pl., Natale Bonifacio (Prospero Parisio); n.i.* TOOLEY, *NL 4F341.*

28. Negro's Head with Headband A; 42 mm tall; thick neck (splayed at base). A4/6(p10)

1557; *[Siege of St. Quentin 10-27 August 1557]; 23.5 x 44 cm; n.pl., n.pr.; n.i.* TOOLEY, *NL 2F29.*

29. Negro's Head with Headband B; 42 mm tall; one thick ribbon; thin neck. A4/6(p10)

1565; [Gulf of Venice]; n.pl.,n.pr.; n.i. Tooley, NL 2F150, later state.

1570; [Sicily]; n.pl.,n.pr.; Tooley 519, NL 4F351.

***1570**; [Battle of Tunis]; n.pl., n.pr.; Tooley 260, KN 87-HEN-14.*

30. Negro's Head with Headband C; 43 mm tall; thick neck but unsplayed at base; thin ribbons (lower one ending in triangle). A4/6(p10) Other manuals: Beans 28 (possible)

1564; [Lombardy]; Rome, Antonio Lafreri; Tooley 352, GHB 221.

1564; [Lombardy]; Rome, Antonio Lafreri; Tooley 352, NL 4F305.

1567; [Messina]; Rome, Antonio Lafreri (Gasparo Algaria); Tooley 388, NL 4F227r (marginal strip).

1569; [Poitiers]; n.pl., n.pr.; Tooley 452, NL 2F51.

1569; [Poitiers]; n.pl., n.pr.; Tooley 614, NL 2F52.

***1570**; Tunise. Castello Vecchio Fortezza della Goletta; 37 x 50 cm; n.pl., n.pr.; n.i. Tooley, NL 2F56.*

n.d. [Sicily]; n.pl., n.pr.; Tooley 519, NC 2:130:43.

29

30

31

31. Negro's Head with Headband in Circle; circle 47 mm dia; above is an 'M', below (on either side of the central wire) two 'G's, all three characters being single wire. A4/6(p10)[U1]

n.d. *Magini. Geografia moderna di tutta l'Italia; Giacomo Gastaldi, n.i.* *Tooley, NL 4F198.*

32

32. Angel; 41 mm tall. A5/2/3 Other manuals: Beans 40 (sim.)

1570; [*Greece*]; *Rome, Antonio Lafreri; Tooley 291, NL 2F186.*

n.d. [Folio sheet of Great Britain place names]; n.pl., n.pr.; Tooley 275, NL 4F130 (slight distortion of chain lines).

33. Justice holding sword and scales, with base formed by two lion heads (56 x 69 mm). B4/3/1[(r:p05M14)-(l:p05N28/2)]

1566; *[North America – New France]; Venice, Bolognino Zaltieri [Paolo Forlani]; TOOLEY 80, NL VCA 15.*

1566; [France]; Venice, Bolognino Zaltieri (Paolo Forlani); TOOLEY 211, NL VCA 16r.

34. Bear with collar and tongue hanging out (passant). C2. Other manuals: BRIQUET 12382 (sim.)

1571; *[German map of the Eastern Mediterranean (battle of Lepanto)]; n.pl., "B. X."; n.i. TOOLEY, NL 2F59.*

33

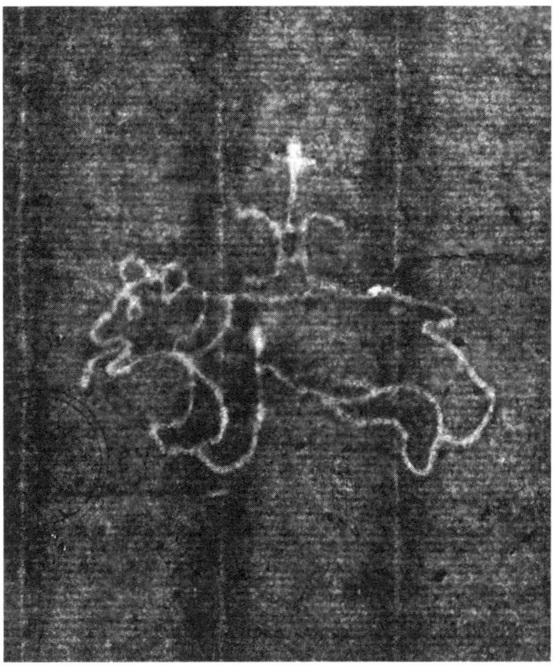

34

35

36

35. Horse passant in circle under star A (nape of neck touching center chain line) chain lines 27-28 mm. C14[U1-J5/4]

1561; [Austria – Alps and Tirol]; Vienna, n.pr.; n.i. TOOLEY, Doria Atlas 52 (ms).

1562; [Poland]; Venice, Giacomo Gastaldi; TOOLEY 453, NL 2F90.

***1562**; [Poland]; Venice, Giacomo Gastaldi; TOOLEY 453, NL 2F96.*

1565; [Ancona]; Venice, Ferrando Bertelli; TOOLEY 104, Doria Atlas 66 (ms).

1566; [World map from Oronce Finé]; Venice, G. Cimerlino; TOOLEY 19, Doria Atlas 1.

1567; [Europe]; Venice, Girolamo Olgiati; TOOLEY 31, Doria Atlas 8a.

1567; [Duchy of Milan]; Venice, Ferrando Bertelli; TOOLEY 389, Doria Atlas 58.

1570; probable date of Doria Atlas marginal strips.

36. Horse passant in circle under star B (nape of neck 2 mm from center chain line) chain lines 27-28 mm. C14[U1-J5/4]

1545; [Austria and Upper Rhine]; Vienna; n.i. TOOLEY, Doria Atlas 50 (ms).

1548; [Palestine]; n.pl., Antonio Salamanca; TOOLEY 433, GHB 210.

***1562**. Il disegno della geografia moderna del Regno di Polonia; [Venice] Fabio Licinio; part of TOOLEY 453, NL 2F97.*

1563; [Rome]; Venice, Ferrando Bertelli; TOOLEY 480, Doria Atlas 69 (ms).

1563; [Duchy of Oswieczin and Zator]; Venice, n.pr.; TOOLEY 427, Doria Atlas 40 (ms).

1567; [Europe]; Venice, Girolamo Olgiati; TOOLEY 31, Doria Atlas 8b.

n.d. [Sopoto]; Rome, Antonio Lafreri; TOOLEY 525, Boswell 240.

1570; probable date of Doria Atlas marginal strips.

37. Lion passant A. C17/1. Other manuals: BEANS 2 (sim.); BRIQUET 10530 (sim.)

1552; *[Germany]; Venice, Giacomo Gastaldi (Enea Vico, Gabriel Giolito); TOOLEY 251, Boswell 219.*

1552; [Germany]; Venice, Giacomo Gastaldi (Enea Vico); TOOLEY 250, Doria Atlas 32.

38. Lion passant B. C17/1

1538. *Isola di Cipro; Venice, Matteo Pagano; TOOLEY 179; NC 1:48:26 left margin.*

1538. Isola di Cipro; Venice, Matteo Pagano; TOOLEY 179; NC 1:48:26 center margin.

1538. Isola di Cipro; Venice, Matteo Pagano; TOOLEY 179; NC 1:48:26 right margin.

37

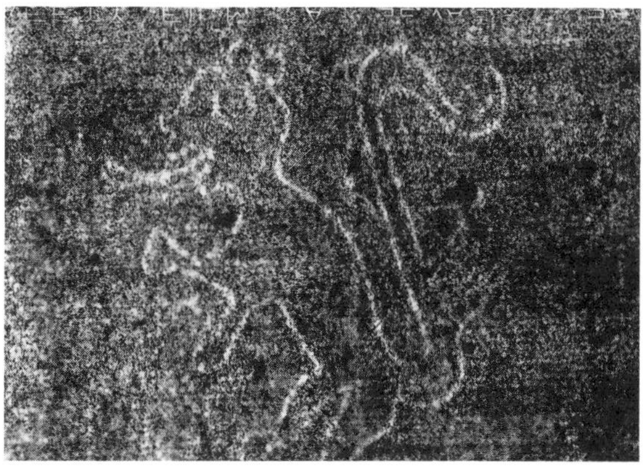

38

39

40

39. Lion rampant in oval (38 x 62 mm). C17/1[U1]

***1590* ca**; *[Messina]; n.pl., Matteo Florimi; n.i. TOOLEY, VAN DER HEYDEN 29; NL 4F226.*

40. Lion rampant in oval surmounted with crown A; oval 43 x 61 mm; C17/1[U1]-R3. Other manuals: HEAWOOD CAT 3130 (sim.)

***1590* ca**; *[Pavia]; n.pl., Matteo Florimi; n.i. TOOLEY, VAN DER HEYDEN 37; NL 4F240.*

n.d. [Gerbi]; n.pl., n.pr.; TOOLEY 244, NL 2F240.

41. Lion rampant in oval surmounted with crown B; oval (39 x 60 mm) C17/1[U1]-R3. Other manuals: HEAWOOD CAT 3131 (sim.)

n.d. *[View of Siena from west]; n.pl., Matteo Florimi; n.i.* TOOLEY, VAN DER HEYDEN *43 or 44; CSUF EFEZX65277.*

42. Lion of St. Mark A. C17/3. Other manuals: ZONGHI 1753 (sim.)

1554*; La vera discrittione di tutta la Francia, la Spagna et la Flandra; n.pl., n.pr.; n.i.* TOOLEY, *NL 2F75.*

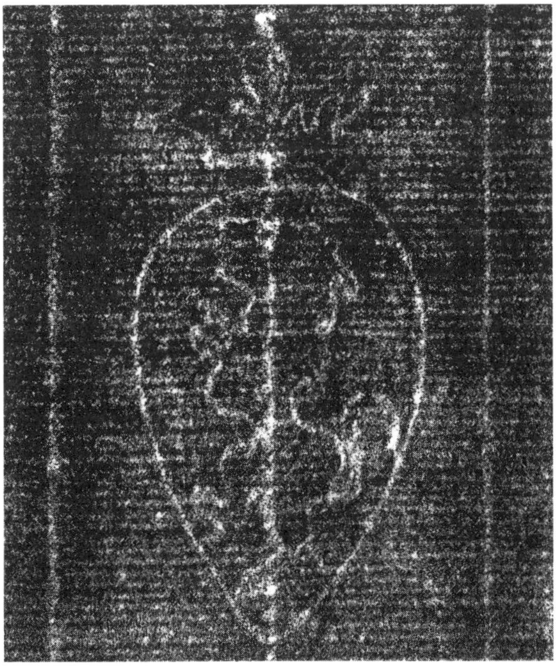

41

42

43

44

43. Lion of St. Mark B. C17/3

1563; [*Oswieczin*]; *Venice, Stanislas Porinski; Tooley 427, NL VCA26.*

44. Lion of St. Mark C. C17/3. Other manuals: Beans 77 (somewhat sim.)

1566; [*Piedmont*]; *Venice, Giacomo Gastaldi; Tooley 450, NL 4F315.*

45. Lion of St. Mark D, inside oval shield (35 x 59 mm) with cardinal's hat. C17/3[M13]

***1580** ca; [Siena]; n.pl., Orlando Malavolta; n.i. Tooley, NL 4F277.*

1593; [Rome]; n.pl., Antonio Tempesta; n.i. Tooley, NL 4F256.

46. Lamb, Paschal with bent standard A, chain lines 32 mm. C19/2[T2/1]

***1559**; [Hungary]; Rome, n.pr.; Tooley 307, NL 4F181.*

1560; [Central Europe]; Rome, Antonio Lafreri; Tooley 26, NL 4F118.

1560; [Central Europe]; Rome, Antonio Lafreri; Tooley 26, NL 4F118v.

45

46

47

48

47. Lamb, Paschal with straight standard B, chain lines 34 mm. C19/2[T2/1]

1562; [World]; n.pl., n.pr.; TOOLEY 24, NL 2F3.

1570; *[Dalmatia]; n.pl., n.pr.; TOOLEY 600, NL 2F204.*

48. Lamb, Paschal with straight standard C, chain lines 23 mm. C19/2[T2/1]

1564; [Rome (environs and Latium]; Rome; TOOLEY 481; GHB 217.

n.d. *[Corfu]; n.pl., n.pr.; TOOLEY 164, NL 2F189.*

49. Lamb, Paschal with straight standard D, chain lines 32 mm. C19/2[T2/1]

1566; *[Brabant]; Venice, Michele Tramezzino (Venice, Jacob Bos); Tooley 221, NL 4F142.*

n.d. [World, double cordiform]; Rome, Antonio Lafreri; Tooley 2, NC 2:130:4.

50. Lamb, Paschal with straight standard E, chain lines 27 mm. C19/2 [T2/1]. Other manuals: Beans 26 (sim.); Massari (1983) 83; Roberts – Lamb B (sim.)

1557; *[St. Quentin]; n.pl., n.pr.; Tooley 506, GHB 240.*

1558; [Belgium]; n.pl., n.pr.; Tooley 131, NL 4F148.

49

50

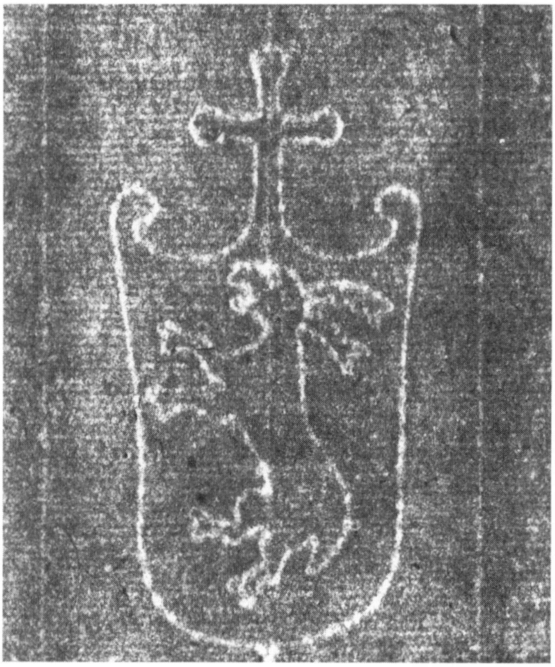

51

52

51. Deer in shield under cross A, with finials. C20[M13-S1]. Other manuals: ZONGHI 899 (sim.); BEANS 21 (sim.)

1567; [Apulia]; Venice, Giacomo Gastaldi; TOOLEY 114, NL 2F173.

1590 ca; [Pozzuoli eruption 29 Sept. 1538]; n.pl., Petri de Nobilibus; TOOLEY 458, NL 4F344 (upper part missing).

1532; *[Austria]; [Rome].,Agostino Veneziano (Antonio Salamanca); n.i. TOOLEY, NL 2F25.*

n.d. [Thionville]; n.pl., n.pr.; TOOLEY 548, NL 4F99 (ms).

52. Deer in shield under cross B, with finials. C20[M13-S1]. Other manuals: ZONGHI 904 (sim.)

1561; [Lombardy]; Venice, Paolo Forlani; TOOLEY 350, Boswell 254.

1562; [Italy]; Venice, Ferrando Bertelli; TOOLEY 333, Doria Atlas 55.

1567; [Milan]; Venice, Ferrando Bertelli; TOOLEY 389, NL 2F159.

n.d. [Thionville]; n.pl., n.pr.; TOOLEY 548, NL 4F99.

n.d. *[Algiers]; n.pl., n.pr.; TOOLEY 97, NL 2F239r.*

53. Deer in shield under cross C, without finials. C20[M13-S1]. Other manuals: HEAWOOD 1929 18 (sim.)

1566*; [France]; Venice, Bolognino Zaltieri; TOOLEY 211, NL VCA 16v.*

54. Eagle displayed in circle under crown A; circle 40 x 42 mm; crown and trailing wing touching circle; tail 4 mm from circle. D5/1[U1-R3]. Other manuals: BRIQUET 209 (sim.); BEANS 59 (somewhat sim.)

1559; [Spain]; Venice, Nicolò Nelli; TOOLEY 501, Boswell 187v.

1564*; [Ancona]; Rome, Vincenzo Lucchino; TOOLEY 102, Boswell 198r.*

1567; [Messina]; Rome, Antonio Lafreri (Gasparo Algaria); TOOLEY 388, Boswell 232r.

1570; [Lombardy]; Rome, Giacomo Gastaldi (Antonio Lafreri); TOOLEY 354, Boswell 194v.

1570; [Germany]; n.pl., Claudio Duchetti (Giacomo Gastaldi); TOOLEY 257, Boswell 178r.

1574; [Tunisia]; n.pl., n.pr.; TOOLEY 563, NL 2F65.

1580; [Palermo]; Rome, Claudio Duchetti; TOOLEY 431, Boswell 234r.

1582; [Italy]; Rome, Antonio Lafreri (Gasparo Algaria); TOOLEY 338, Boswell 192r.

1585; [Ancona]; Rome, Claudio Duchetti; TOOLEY 109, NL 2F115.

1586; [Campania]; Rome, Claudio Duchetti (Ambrogio Brambilla); TOOLEY 460, Boswell 226r.

1587; [Abruzzi]; Rome, Natale Bonifacio (Nicolas van Aelst); TOOLEY 94, Boswell 199r.

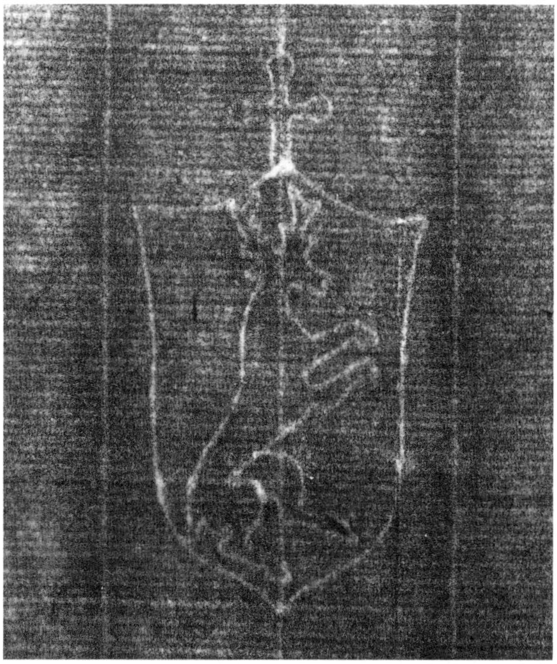

53

54

55

56

55. Eagle displayed in circle under crown B; circle 47 x 48 mm; crown 2 mm from circle; tail 6 mm from circle. D5/1[U1-R3]. Other manuals: BRIQUET 209 (sim.)

1558; [Calais]; n.pl., n.pr.; TOOLEY 152, NL 4F97.

1562; *[Poland]; Venice, Giacomo Gastaldi; TOOLEY 453, GHB 145.*

1564; [Lombardy]; Rome, Antonio Lafreri; TOOLEY 352, NL 4F306.

1565; [Marca di Ancona]; Venice, Ferrando Bertelli; TOOLEY 104, NL 4F325.

1565; [Malta]; Rome, Antonio Lafreri; TOOLEY 374, NL 2F39.

1590 ca; [Pozzuoli eruption 29 Sept. 1538]; n.pl., Petri de Nobilibus; TOOLEY 458, NL 4F344.

n.d. [Sardinia]; n.pl., n.pr.; TOOLEY 509, NL 2F178.

n.d. [Corsica]; n.pl., Fabio Licinio; TOOLEY 169, NL 4F81.

56. Eagle displayed in circle under crown C; circle 47 x 49 mm; crown 3 mm from circle; tail 7 mm from circle; "thin" outside feather on trailing wing. D5/1[U1-R3]. Other manuals: BRIQUET 209 (sim.)

1554; Enea Vico; n.i. TOOLEY, NL 4F132.

1559; [Hungary]; Rome, Lazius (1556 orig.); TOOLEY 307, Doria Atlas 42 (later state with bent circle).

1561; [Asia, third part]; Venice, Giacomo Gastaldi; TOOLEY 61, NL 6F40.

1561; [Asia, third part]; Venice, Giacomo Gastaldi; TOOLEY 61, NL 6F40.

1563; *[Malta]; Rome, n.pr.; TOOLEY 375, NL 4F352.*

1565; [Malta]; Rome, Antonio Lafreri; TOOLEY 384, NL 2F40.

1565; [Malta]; Rome, Antonio Lafreri; TOOLEY 384, NL 2F40ms.

1565; [Malta]; Rome, Antonio Lafreri; TOOLEY 360, NL 2F42.

n.d. [Thionville]; n.pl., n.pr.; TOOLEY 548, NL 4F99.

n.d. [Austria and Hungary]; n.pl., n.pr.; TOOLEY 121, NL 4F175 (ms).

n.d. [Greece, Gulf of Artha]; n.pl., n.pr.; TOOLEY 290, NL 2F23.

n.d. [Elba]; n.pl., n.pr.; TOOLEY 196, NL 2F111.

57. Eagle displayed in circle under crown D; circle 46 x 50 mm (crooked); tail 3 mm from circle. D5/1{U1-R3]. Other manuals: BRIQUET 209 (sim.)

1558; *[France]; Rome, Pirro Ligorio (Sebastiano di Re, Michele Tramezzino); TOOLEY 208, Boswell 190r.*

1570; [World]; Venice, Paolo Forlani (Claudio Duchetti); TOOLEY 15, Boswell 176v.

1570; [Lombardy]; Rome, Giacomo Gastaldi (Antonio Lafreri); TOOLEY 354, Boswell 195r.

58. Eagle displayed in circle under crown E; circle 42 x 45 mm; crown 1 mm from circle; tail 2 mm from circle. D5/1[U1-R3]. Other manuals: BRIQUET 209 (sim.)

1570; [World]; Venice, Paolo Forlani (Claudio Duchetti); TOOLEY 15, Boswell 177r.

1582; [Rome]; Rome, Claudio Duchetti; TOOLEY 494, NL 2F131.

1585 ca; *[Florence]; n.pl., Claudio Duchetti; TOOLEY 206, Boswell 236r.*

1590; [Rome]; Rome, Ambrogio Brambilla (Claudio Duchetti); TOOLEY 497, Boswell 229r.

57

58

59

59. Eagle displayed in circle under crown F; circle 47 x 47 mm; crown (convex to and 2 mm from circle); tail 2 mm from circle. D5/1[U1-R3]. Other manuals: BRIQUET 209 (sim.)

__1602?__; Ferrara. Dedicata al Cardinale S. Giorgio; [Rome], n.pr.; n.i. TOOLEY, NL 4F218.

60

60. Eagle displayed in circle under crown G; circle 42 x 43 mm; crown (crooked 1-3 mm from circle); tail 5 mm from circle. D5/1[U1-R3]. Other manuals: BRIQUET 209 (sim.)

1559; [Spain]; Rome, Vincenzo Lucchino; TOOLEY 530, Boswell 188r.

__1564__; [Lazio]; Rome, Claudio Duchetti; TOOLEY 482, Boswell 200r.

1585 ca; [Campania, Puglia]; Rome, Claudio Duchetti; TOOLEY 405, Boswell 201r.

61. Eagle displayed in circle under crown H; circle 43 x 46 mm; crown 2 mm from circle; tail 4 mm from circle. D5/1[U1-R3]. Other manuals: BRIQUET 209 (sim.)

1555; [Rome]; Rome, Antonio Lafreri (Jacob Bos); TOOLEY 486, CSUF ac 1333.
1555; [Switzerland]; Rome, Antonio Lafreri (Jacob Bos orig.); TOOLEY 534, NC 2:130:32 left.
1555; [Switzerland]; Rome, Antonio Lafreri (Jacob Bos orig.); TOOLEY 534, NC 2:130:32 right.

62. Eagle displayed in circle under crown I; circle 44 x 45 mm; crown 4 mm from circle, crown points 28 mm apart; tail 2 mm from circle. D5/1[U1-R3]. Other manuals: BEANS 44 (sim.); BRIQUET 209 (sim.)

***1580**; [Lisbon]; Rome, Mario Cartaro; TOOLEY 345, GHB 365.*

61

62

63

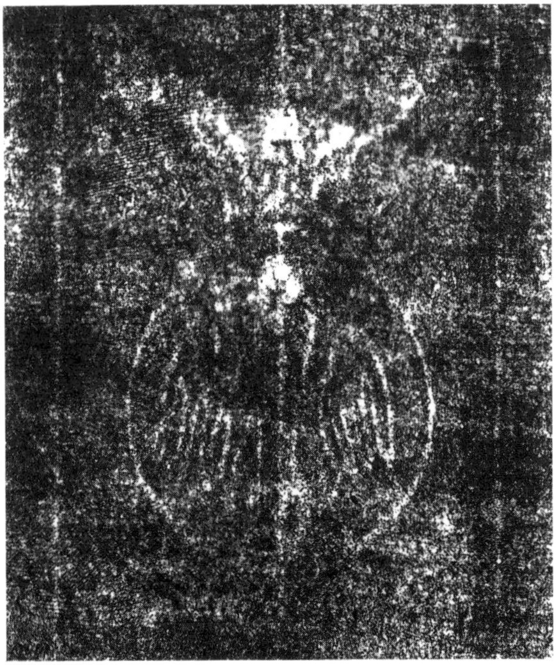

64

63. Eagle displayed in circle under crown J; circle 43 x 45 mm; crown 4 mm from circle; couped tail 9 mm from circle. D5/1[U1-R3]. Other manuals: BRIQUET 209 (sim.)

n.d. *[Siena]; n.pl., Stefano Buonsignori (Don Vitus); n.i. TOOLEY, GHB 157.*

64. Eagle displayed in circle under crown K; circle 41 x 46 mm; crown 2 mm from circle. D5/1[U1-R3]. Other manuals: BRIQUET 209 (sim.)

1592 ca*; [Geneva]; Rome, Aliteni Gatti; TOOLEY 239, Boswell 242r.*

65. Eagle displayed in circle under crown L; circle 36 x 39 mm, crooked; crown 2 mm from circle; tail 4 mm from circle. D5/1[U1-R3]. Other manuals: BRIQUET 209 (sim.)

n.d. *Piacenza. La nobilissima città di Piacenza; [Siena]; Matteo Florimi; n.i. TOOLEY, VAN DER HEYDEN 39, NL 2F125.*

66. Eagle displayed in circle under crown M; circle 41 x 44 mm; crown 1 mm from circle; tail 4 mm from circle; 2 chainlines 63 mm. D5/1[U1-R3]. Other manuals: BRIQUET 209 (sim.)

1580 ca *Origine et translazione della Chiesa di Santa Maria di Loreto; n.pl., n.pr.; n.i. TOOLEY, NL 2F120.*

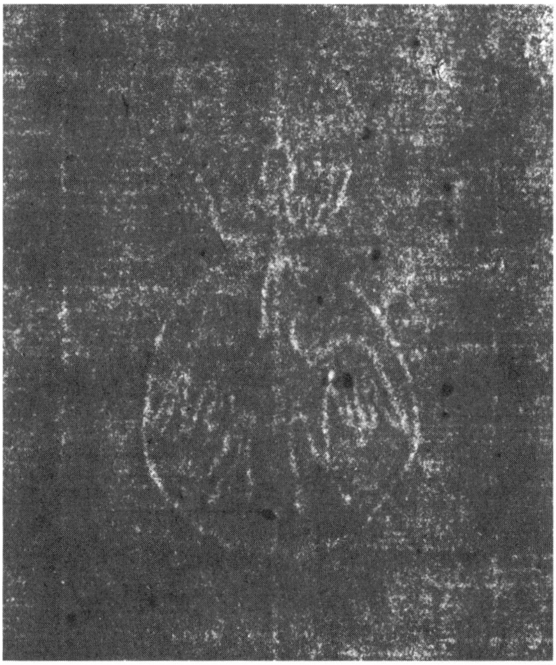

65

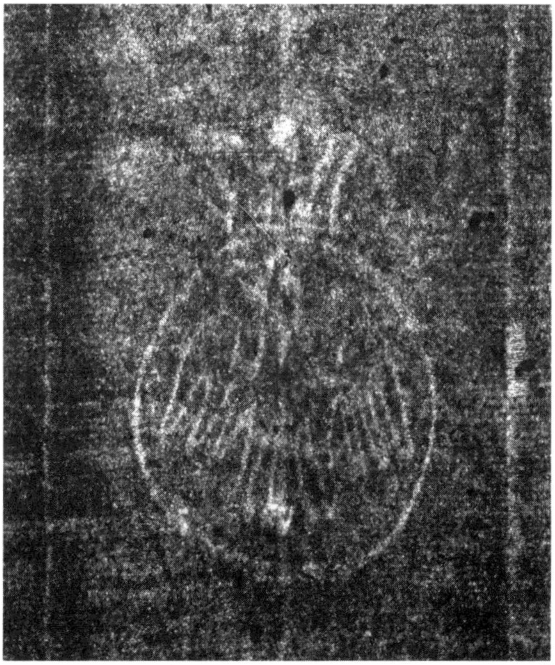

66

67

68

67. Eagle displayed in circle under crown N; circle 44 x 44 mm; crown crooked; crown points 24 mm apart; tail 3 mm from circle. D5/1[U1-R3]. Other manuals: BRIQUET 209 (sim.)

1555; *[Rome]; Rome, Antonio Lafreri (Jacob Bos); TOOLEY 486, CSUF 1333v.*

n.d. [Pozzuoli eruption 29 Sept. 1538]; Rome, Antonio Lafreri (G.A. orig.); TOOLEY 457, NC 2:130:72.

68. Eagle displayed in circle under crown O; circle 42 x 45 mm, crooked; crown 2 mm from circle; tail 2 mm from circle. D5/1[U1-R3]. Other manuals: BRIQUET 209 (sim.)

n.d. *[Mantova]; n.pl. [Siena]., Matteo Florimi; n.i. TOOLEY, VAN DER HEYDEN 27, NL 4F225.*

69. Eagle displayed in circle under crown P; circle 42 x 45 mm; crown 1 mm from circle; tail 2 mm from circle; 2 chainlines 57 mm. D5/1[U1-R3]. Other manuals: BRIQUET 209 (sim.)

1589; *Urbis Romae novissima delineato MDLXXXVIII; Rome, Giovanni Battista de Cavalleri; n.i.* *TOOLEY, GHB 402.*

70. Eagle displayed in circle under crown Q; circle 47 x 49 mm; crown 3mm from circle; tail 4 mm from circle. D5/1[U1-R3]. Other manuals: BRIQUET 209 (sim.)

1558; [*Naples*]; *Rome, Sebastianus Clodiensis (Pirro Ligorio); TOOLEY 403, NL 4F339.*

69

70

71

72

71. Eagle displayed in circle under crown R; circle 40 x 43 mm; three-pointed crown touching circle; tail 4 mm from circle. D5/1[U1-R3]. Other manuals: BRIQUET 209 (sim.)

1565; *Dissegno del Tornamento il lune di Carnevale in Roma nel Teatro Vaticano; [Rome], Antonio Lafreri (Etienne Dupérac); n.i.* TOOLEY, *NL 4F265.*

72. Eagle displayed in circle under crown S; circle 42 x 43 mm; five-point crown 3 mm from circle with all vertical points; tail 2 mm from circle. D5/1[U1-R3]. Other manuals: BRIQUET 209 (sim.)

1590 ca; *[Rome]; n.pl., Matteo Florimi; n.i.* TOOLEY, VAN DER HEYDEN *41, NL 4F254.*

73. Eagle displayed in circle under crown T; circle 41 x 43 mm; five-point crown 2 mm from circle; tail 4 mm from circle. D5/1[U1-R3]. Other manuals: BRIQUET 209 (sim.)

1590 ca; [Pisa]; n.pl., Matteo Florimi; n.i. TOOLEY, VAN DER HEYDEN 40, NL 4F245.

1590 ca; *[Naples]; Siena, Matteo Florimi; n.i. TOOLEY, VAN DER HEYDEN 32, NL 4F234.*

74. Eagle displayed in circle under crown U; circle 41 x 43 mm; crown 3 mm from circle; tail 3 mm from circle. D5/1[U1-R3]. Other manuals: BRIQUET 209 (sim.)

***1590** ca; [Palermo]; Siena, Matteo Florimi; n.i. TOOLEY, VAN DER HEYDEN 34, NL 4F239.*

73

74

75

76

75. Eagle displayed in circle under crown V; circle 46 x 49 mm; crown convex to and 2 mm from circle; tail 1 mm from circle. D5/1[U1-R3]. Other manuals: BRIQUET 209 (sim.)

1602; *Palatii Farnese Romae; Rome, Giovanni Orlandi (Claudio Duchetti); n.i.* TOOLEY, *NL 4F276.*

76. Eagle displayed over two turrets. Total height 71 mm. D5/1[2*(K7/1)]. Other manuals: MUSEO COMUNALE DI GUBBIO 3 (sim.)

1569; [*Orvieto*]; *n.pl., n.pr.;* TOOLEY *424, NL 4F235.*

77. Goose in Circle A; circle 39 x 40 mm. D7[U1]

n.d. *[Corfu]; n.pl., n.pr.; n.i.* TOOLEY, *NL 2F190.*

78. Goose in Circle B; circle 46 x 50 mm; undulating central chain line. D7[U1]

1590 ca*; [Acquapendente]; Rome, Paolo Gratiano (Petri de Nobilibus);* TOOLEY *95, Boswell 244r.*

77

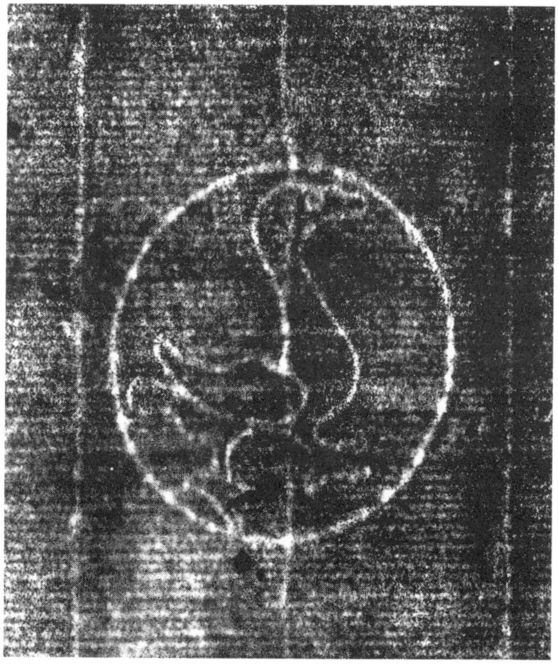

78

79

79. Goose in Circle C; circle 46 x 47 mm; undulating central chain line. D7[U1]

***1540* ca**; [*Rome*]; *n.pl., Antonio Salamanca; n.i.* *Tooley, NL 4F259.*

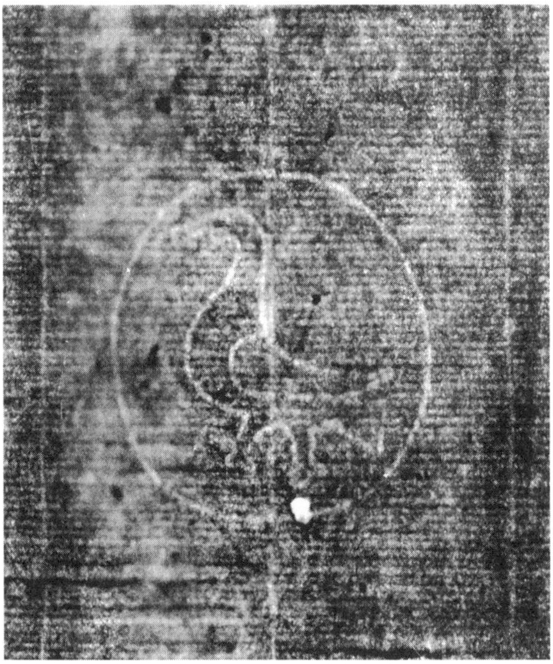

80

80. Goose in Circle D; circle 43 x 45 mm. D7[U1]

1590; [*Piedmont*]; *Venice, Francesco Vallegio; n.i. Tooley, NL 2F164 (sheet f).*

1590; [Piedmont]; Venice, Francesco Vallegio; n.i. Tooley, NL 2F164 (sheet c) (ms).

81. Goose in Circle Under Cross A; circle 42 x 46 mm; saltire cross, variant A (back of head touches central chainline). D7[U1-S1]

1590; *[Piedmont]; Venice, Francesco Vallegio; n.i. Tooley, NL 2F164 (sheet d).*

1590; [Piedmont]; Venice, Francesco Vallegio; n.i. Tooley, NL 2F164 (sheet b).

82. Goose in Circle Under Cross B; circle 42 x 46 mm; saltire cross, variant B (back of head away from central chainline). D7[U1-S1]

1590; [Piedmont]; n.pl., Francesco Vallegio; n.i. Tooley, NL 2F164 (sheet a).

1590; *[Piedmont]; n.pl., Francesco Vallegio; n.i. Tooley, NL 2F164 (sheet d).*

81

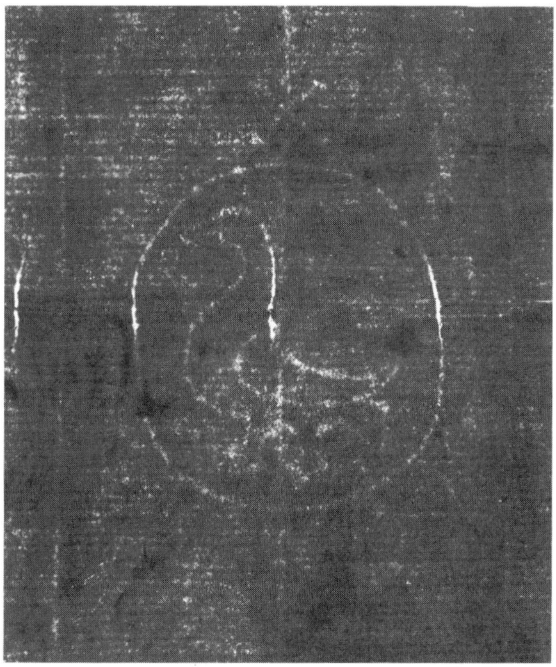

82

83

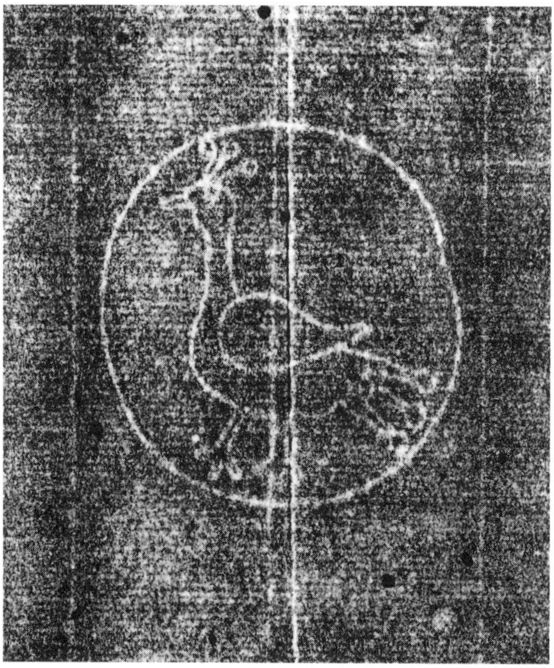

84

83. Goose over mountains in two circles; outer circle 50 x 52 mm. D7-J3/3[2*(U1)]

1564; [Rome]; Rome, Claudio Duchetti; Tooley 481, NL 4F328.

84. Peacock in circle A; circle 50 mm dia with dent; crest away from central chainline. D9<f:U1>. Other manuals: Beans 67 (sim.)

1560; [North Atlantic]; Venice, Paolo Forlani (Giovanni Francesco Camocio); Tooley 77, NL 4F414.

1560 ca; [Piedmont]; Rome, Antonio Lafreri; Tooley 446, NL 4F313.

n.d. [Corsica]; n.pl., Fabio Licinio; Tooley 169, NL 2F79.

n.d. [Gerbi]; n.pl., n.pr.; Tooley 244, NL 2F240.

n.d. [Greece]; n.pl., Franciscus Salamanca (Sebastianus Clodiensis); Tooley 284, Boswell 218r.

n.d. *[Crete]; n.pl., n.pr.; Tooley 177, Boswell 224.*

85. Peacock in circle B; circle 50 mm dia; crest touches central chainline. D9<f:U1>. Other manuals: BEANS 67 (sim.); BORRONI SALVADORI 74 (sim.)

1560 ca; Creta insula hodie Ca[n]dia; [Rome], Antonio Lafreri; n.i. TOOLEY, NL 2F198.

1591 *ca; [Greece]; n.pl., Sebastianus Clodiensis; TOOLEY 284, NL 4F357.*

n.d. [Cyprus]; n.pl., n.pr.; TOOLEY 181, NL 2F224.

n.d. [Greece]; n.pl., Franciscus Salamanca (Sebastianus Clodiensis); TOOLEY 284, Boswell 218v.

n.d. [Gerbi]; Rome, Antonio Lafreri; TOOLEY 244, NC 2:130:54.

86. Small bird in shield under crown. D12<f:M13{t:R3}>

1571*; Das is die wahrhafte Contrafettung des stat Zara in Dalmatia; n.pl., n.pr.; n.i. TOOLEY, NL 2F60.*

85

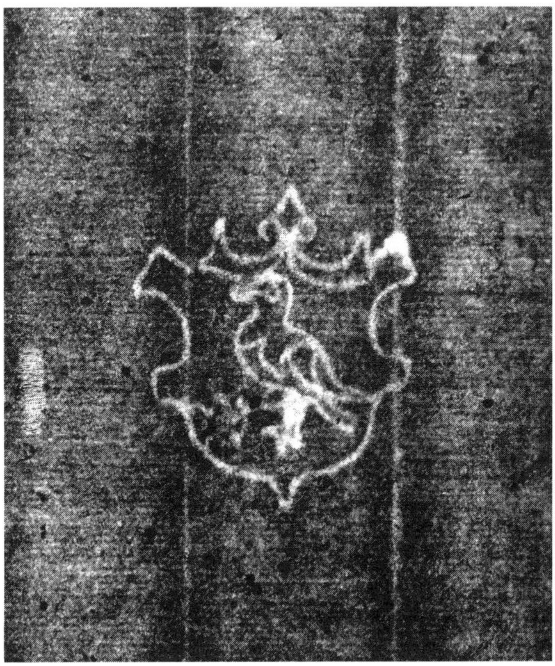

86

87

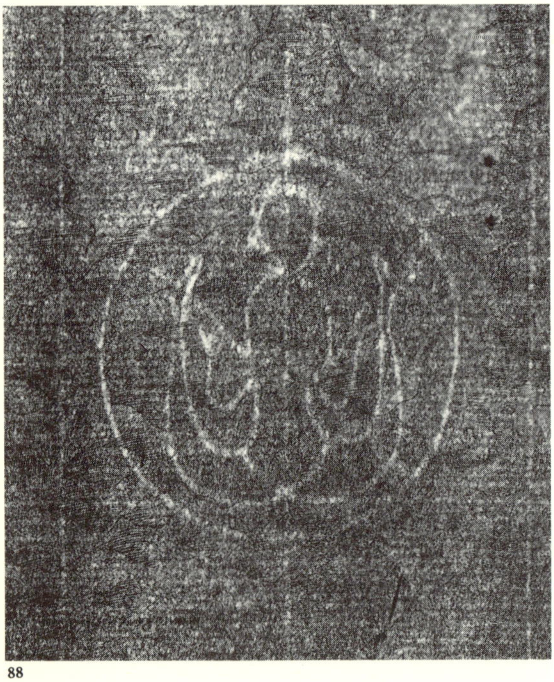

88

87. Siren holding two crosses 48 x 53 mm. F3/2 Other manuals: Beans 7 (sim.)

1556; *[Great Britain]; Rome, n.pr.; Tooley 269, CSUF 892.*

88. Siren in circle A; circle 49 x 51 mm; sharp curve in one side of neck. F3/2[U1]

1565; [Italy]; Venice, Ferrando Bertelli; Tooley 333, NL 4F202v.

1566; [Ireland]; Venice, Bolognino Zaltieri; Tooley 316, CSUF ac 1073.

1567; [Brabant]; Venice, Girolamo Olgiati (Jacob van Deventer); Tooley 146, GHB 121.

1567; [Flanders]; Venice, Paolo Forlani; Tooley 203, NL 4F151.

1570 ca; *[Sicily]; n.pl., Ferrando Bertelli; Tooley 521, NL 4F350.*

89. Siren in circle B; circle 50 x 49 mm; smooth neck. F3/2[U1] Other manuals: BEANS 34 (sim.)

1562; *[Savoy]; Venice, Ferrando Bertelli (Forlani); TOOLEY 513, NL 2F77.*

1563; [Tuscany]; Venice, Paolo Forlani (Ferrando Bertelli); TOOLEY 574, NL 4F335.

1565; [Italy]; Venice, Ferrando Bertelli; TOOLEY 333, NL 4F202r.

1566; [Switzerland]; Vincenzo Lucchino; TOOLEY 538, NL 4F163v.

1567; [Belgium]; Girolamo Olgiati; TOOLEY 146, NL 4F153.

n.d. [Belgium]; n.pl., n.pr.; TOOLEY 136, NL 2F82.

n.d. [Low Countries]; Venice, Paolo Forlani; TOOLEY 136, GHB 264.

90. Siren in circle C; one side of circle flattened slightly. F3/2[U1]

1590 ca; *[Perugia]; n.pl. [Siena], Matteo Florimi; n.i. TOOLEY, NL 4F242.*

89

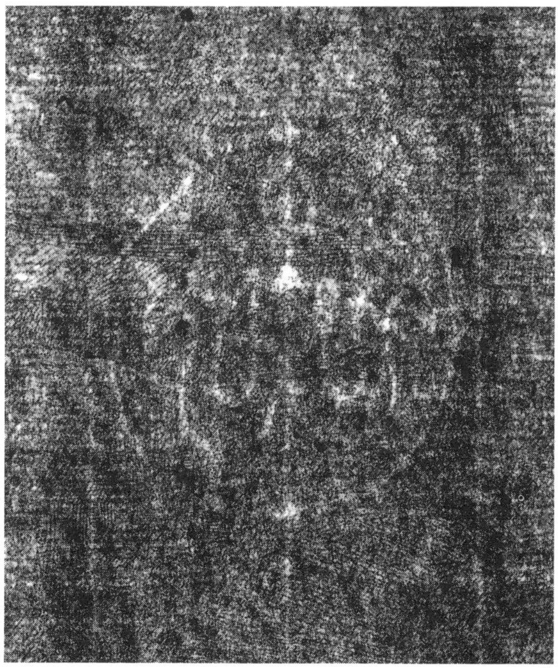

90

91

91. Siren in circle under six-pointed star A; circle 49 x 51 mm; thick waist; total height 73 mm; referred to as "MARY" in WOODWARD 1987; [U1]-J5/4. Other manuals: BEANS 33; HEAWOOD CAT 3800 (sim.); MASSARI (1983) 136b

1557; [Northern Europe]; Venice, Giovanni Francesco Camocio; TOOLEY 41, NYPL map 27.

1564; [Corfu]; Venice, Ferrando Bertelli; TOOLEY 165, NYPL map 57.

1565 [=1568-70]; [World]; Venice, Ferrando Bertelli [Paolo Forlani]; TOOLEY 10, WOODWARD 1990a 35.04, NMC 22899.

1565; [South Asia]; Venice, Giacomo Gastaldi (Ferdinando Bertelli); TOOLEY 73, KN 096-AEN-15.

1566; [North America – New France]; Venice, Bolognino Zaltieri [Paolo Forlani]; TOOLEY 80, NL 2F13.

1566; [North America – New France]; Venice, Bolognino Zaltieri, TOOLEY 80; NMC 6577.

1566; [Tunis]; Venice, Bolognino Zaltieri (Paolo Forlani); TOOLEY 558, CSUF ac 1332.

1566; [Venice]; Venice, Paolo Forlani (Bolognino Zaltieri); TOOLEY 582, GHB 374r.

1566; [World]; [Venice], Giovanni Paolo Cimerlino; TOOLEY 19, NC 2:130:3 right.

1567; [Holland]; n. pl., Bolognino Zaltieri (Girolamo Olgiati); TOOLEY 302, NC 2:130:19.

1567; [Brabant]; Venice, Jacob van Deventer (Girolamo Olgiati, Bolognino Zaltieri); TOOLEY 145, GHB 216.

1567; [Piedmont]; Venice, Paolo Forlani; TOOLEY 451, NL 2F162.

***1567**; [Netherlands]; n. pl., Girolamo Olgiati; TOOLEY 303, NL 2F17.*

1568; [Poland]; Venice, Paolo Forlani (Giacomo Gastaldi); TOOLEY 454,
GHB 'TOOLEY 454'.

1568; [Rome]; Venice, "Leon Pitor"; n.i. TOOLEY: HÜLSEN 19, SCACCIA SCARAFONI 154, KN 051-21-#6r.

1568; [Rome]; Venice, "Leon Pitor"; n.i. TOOLEY: HÜLSEN 19, SCACCIA SCARAFONI 154, KN 051-21-#6v

1568; [Gulf of Venice]; n. pl., Paolo Forlani; TOOLEY 587, NL 2F151.

1568; [Poland]; Venice, Paolo Forlani; TOOLEY 454; NYPL map 28.

1569; [Greece]; Venice, Giacomo Gastaldi (Domenico Zenoi, Donato Bertelli); TOOLEY 289, GHB 181r.

1569; [Italy]; Venice, Domenico Zenoi (Donato Bertelli); TOOLEY 335, GHB [no number, has star in lower left corner]v.

1570; [Zara and Sebenico]; n. pl., Martino Rota; TOOLEY 601, NL 4F103r.

92. Siren in circle under six-pointed star B; circle 49 x 50 mm; thin waist; total height 74 mm. Referred to as "MARTHA" in WOODWARD 1987; F3/2[U1]-J5/4

1559; [Germany]; Venice, Giacomo Gastaldi (Domenico Zenoi, Donato Bertelli); TOOLEY 253, GHB 258.

1562; [Baltic Sea]; Venice, Giovanni Francesco Camocio; TOOLEY 41, NL 4F117.

1563; [Gelderland]; Venice, Giovanni Francesco Camocio (Paolo Forlani); TOOLEY 234, NC 2:130:18 right.

1563; [France]; Venice, Oronce Fine (Domenico Zenoi); TOOLEY 210, NL 4F134v.

1565 [=1568-70]; [World]; Venice, Ferrando Bertelli [Paolo Forlani]; TOOLEY 10, WOODWARD 1990A 35.04, NMC13294

1566; [Sicily]; n. pl., n. pr.; TOOLEY 518, NL 4F349

1566; [Venice]; Venice, Paolo Forlani (Bolognino Zaltieri); TOOLEY 582, GHB 374v.

1566; [World]; [Venice], Giovanni Paolo Cimerlino; TOOLEY 19, NC 2:130:3 left.

1567; [Brabant]; Venice, Bolognino Zaltieri (Girolamo Olgiati); TOOLEY 145, NC 2:130:17.

1568; [Poland]; Venice, Paolo Forlani (Giacomo Gastaldi); TOOLEY 455, GHB 'TOOLEY 455'.

1568; [Poland]; Venice, Paolo Forlani (Giacomo Gastaldi); TOOLEY 455; NYPL map 29.

1568; [Augsburg]; Venice, Luca Quarinoni; TOOLEY 115, NL 4F172.

1568; [Augsburg]; Venice, Luca Quarinoni; TOOLEY 116, GHB 236.

1569; [Italy]; Venice, Domenico Zenoi (Donato Bertelli); TOOLEY 335,

GHB [no number, has star in lower left corner]v.

1569; [Morea]; n. pl., Giovanni Francesco Camocio; TOOLEY 398, NL 2F191.

1569; [Greece]; Venice, Giacomo Gastaldi (Domenico Zenoi, Donato Bertelli); TOOLEY 289, GHB 181v.

1570; [Morea and Peloponnesus]; n. pl., Claudio Duchetti; TOOLEY 400, NL 2F192.

1570; [Rhodes]; n. pl., Claudio Duchetti; TOOLEY 465, NL 2F202.

1570; [Egypt]; n. pl., [Paolo Forlani]; TOOLEY 194, NL 2F244.

1570; [Bavaria]; [Venice], Paolo Forlani; TOOLEY 130, GHB 160.

92

1570; [Zara]; Venice, Paolo Forlani; TOOLEY 600, GHB 182.

1570; [Egypt]; [Venice], [Paolo Forlani]; TOOLEY 194, GHB 214.

1570; [Constantinople]; n. pl., Claudio Duchetti; TOOLEY 156, GHB 245.

n.d.; [Ireland]; n. pl., n. pr.; TOOLEY 315, NC 2:130:7.

n.d.; [Flanders]; n. pl., Claudio Duchetti; TOOLEY 205, NC 2:130:16 right.

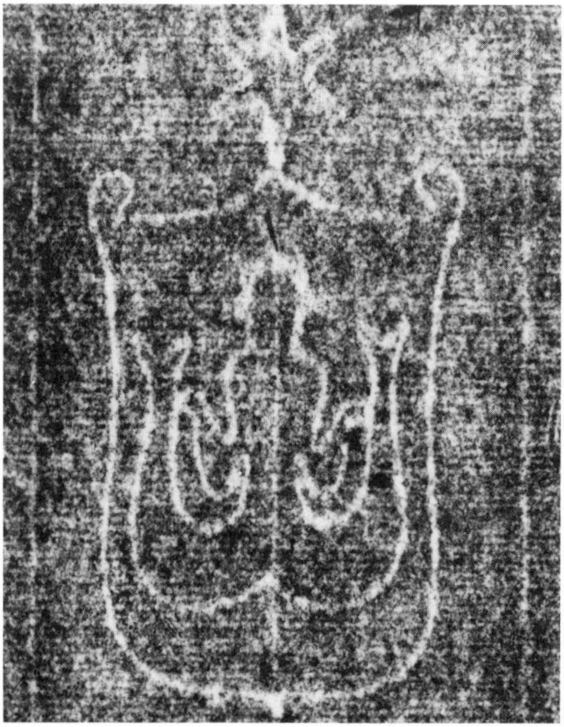

93

94

93. Siren in shield with star A; shield 51 x 71 mm; under six-pointed star; total height 92 mm. F3/2[M13]-J5/4. Other manuals: HEAWOOD CAT 3804 (sim.)

1562; [Northern Europe]; Venice, Giovanni Francesco Camocio; TOOLEY 41, Doria Atlas 35.

1566; *[Switzerland]; Venice, Vincenzo Lucchino; n.i. TOOLEY, GHB 159v.*

1570; [Zara]; n.pl., Martino Rota; TOOLEY 601, Doria Atlas 89.

n.d. [Velez de Gomera]; Rome, Antonio Lafreri (Giovanni Francesco Camocio orig.); TOOLEY 579, NC 2:130:74.

94. Siren in shield with star B; shield 50 x 69 mm; under six-pointed star; total height 91 mm. F3/2[M13]-J5/4 Other manuals: BEANS 36 (sim.)

1563; *[Friuli]; Giovanni Francesco Camocio; TOOLEY 227, NL 2F153.*

1568; [Poland, first part]; Rome, Antonio Lafreri (Paolo Forlani orig.); TOOLEY 454, NC 2:130:22.

n.d. [Palestine]; Donato Bertelli; TOOLEY 439, NL 2F231.

95. Fleur-de-lys A; 40 mm height. G6/2. Other manuals: HEAWOOD CAT 1414 (sim.). Ilkley print.

__1566__; [Danube Basin]; Venice, Paolo Forlani; TOOLEY 29, Doria Atlas 44.

96. Fleur-de-lys B; 39 mm height. G6/2. Other manuals: HEAWOOD CAT 1414 (sim.). Ilkley print.

***n.d.** [Bohemia, 2 sheets]; [Venice], Bolognino Zaltieri; TOOLEY 138, Doria Atlas 47(i).*

n.d. [Bohemia, 2 sheets]; [Venice], Bolognino Zaltieri; TOOLEY 138, Doria Atlas 47(ii).

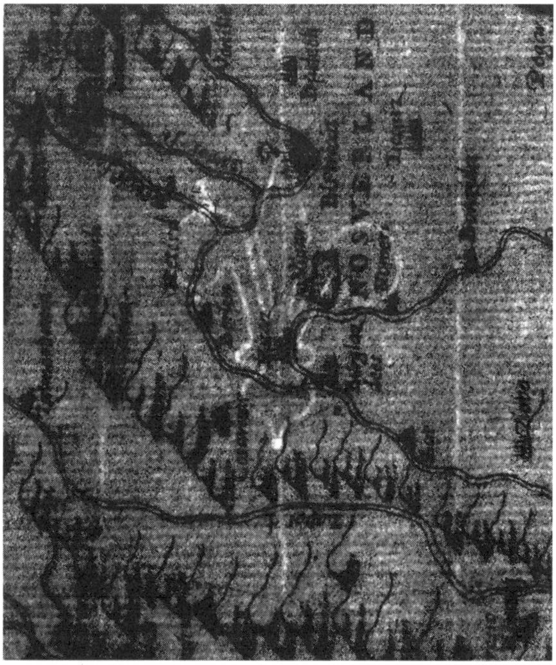

95

96

97

98

97. Fleur-de-lys C; in circle 41 x 42 mm. G6/2[U1]

1566; *[Malta]; Rome, Antonio Lafreri;* TOOLEY *362, GHB 232.*

98. Fleur-de-lys D; in circle 41 x 43 mm. G6/2[U1]

n.d. []; *n.pl., n.pr.; n.i.* TOOLEY, *NL 4F194.*

99. Fleur-de-lys E; in crooked circle 43 x 46 mm. G6/2[U1]

1582; *[Bologna]; Rome, Claudio Duchetti (Antonio Lafreri); Tooley 139, NL 4F214.*

100. Fleur-de-lys F; in circle 45 mm dia; with sprouts. G6/2[U1]. Other manuals: Briquet 7317 (sim.); Borroni Salvadori 56 (sim.)

1558; *Il vero disegno di Thionville; n.pl., n.pr.; n.i. Tooley, GHB 241.*

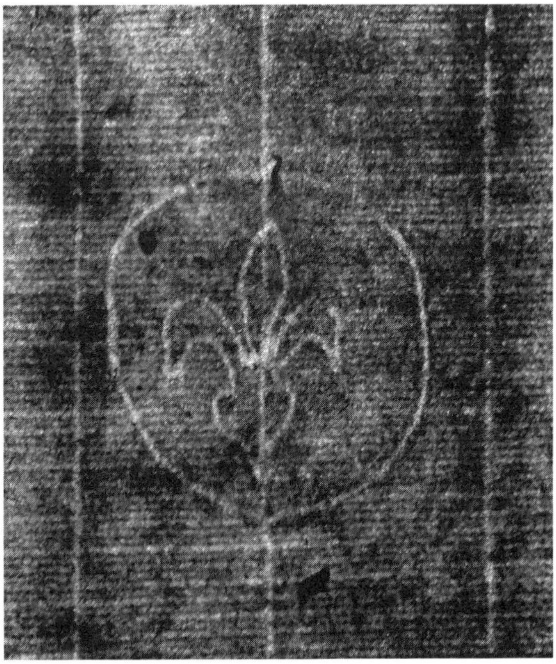

99

100

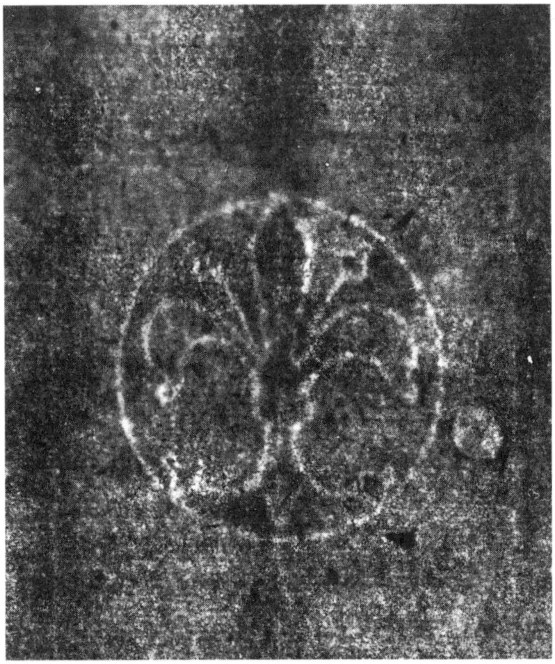

101

102

101. Fleur-de-lys G; in circle 44 x 45 mm; with sprouts; twin of #100?. G6/2[U1]. Other manuals: BRIQUET 7317 (sim.); HEAWOOD 1929 7 (sim.)

1558; [Guines]; Rome, n.pr.; TOOLEY 293, NL 4F98.

1566; *[Malta]; n.pl., Antonio Lafreri; TOOLEY 362, NL 4F353.*

n.d. [Minorca]; Rome, [Antonio Lafreri]; TOOLEY 392,
NC 2:130:12.

102. Fleur-de-lys H; in two concentric circles (41 x 45 mm crooked; 36 mm dia); above triple-mount. G6/2-J3/3 [2*(U1)]

1590 ca; [Milan]; n.pl., Matteo Florimi; n.i. TOOLEY, VAN DER HEYDEN 30, NL 4F231.

1590 *ca;[Tuscany]; Matteo Florimi; n.i. TOOLEY, VAN DER HEYDEN 14, NL 2F168.*

103. Fleur-de-lys in circle under crown A; crooked circle 43 x 45 mm; crown w/ five triangular, uneven points; total height 63 mm. G6/2[U1]-R3

1600; *[Aquila]; n.pl., Jacobus Laurus; n.i.* TOOLEY, *NL 4F213.*

104. Fleur-de-lys in circle under crown B; crooked circle 38 x 42 mm; exaggerated crown with five triangular and uneven points; total height 66 mm. G6/2[U1]-R3

n.d. *[Jerusalem]; Rome, Claudio Duchetti;* TOOLEY *339, NL 4F114.*

103

104

105

106

105. Fleur-de-lys in circle under crown C; circle 39 x 41 mm; five pointed crown: central point in cross-shape, outermost triangular, and intermediate with knobs on their tops; total height 68 mm. G6/2[U1]-R3

***1590** ca; [Piedmont]; n.pl., Luigi Rosaccio; n.i. Tooley, NL 4F316.*

106. Fleur-de-lys in circle under crown D; circle 47 x 50 mm; crown w/five triangular, uneven points; total height 61 mm. G6/2[U1]-R3

***1592**; [Naples]; Rome, Natale Bonifacio; n.i. Tooley, NL 4F342.*

107. Fleur-de-lys in circle under six-pointed star A; circle 41 x 42 mm; total height 67 mm. G6/2[U1]-J5/3

1565; [Malta]; Rome, Antonio Lafreri; TOOLEY 383, CSUF 1831.

1565; *[Malta]; Rome, Antonio Lafreri; TOOLEY 383, NL 2F37.*

108. Fleur-de-lys in circle under six-pointed star B; circle 38 x 35 mm; total height 55 mm. G6/2[U1]-J5/3

1590 *ca; [Italy]; n.pl., Matteo Florimi (Giacomo Gastaldi); n.i. TOOLEY, VAN DER HEYDEN 10, NL 4F207.*

107

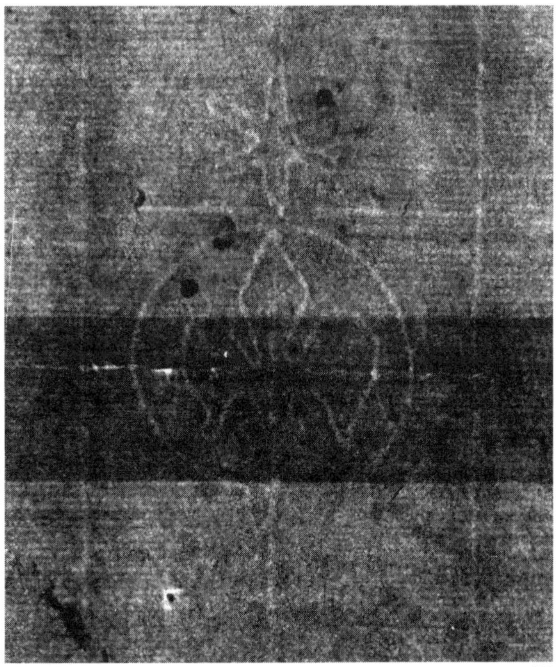

108

109

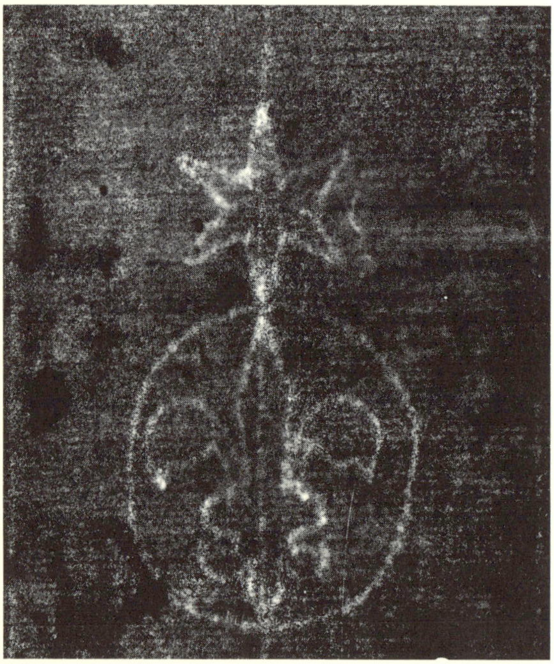

110

109. Fleur-de-lys in circle under six-pointed star C; circle 34 x 34 mm; total height 69 mm. G6/2[U1]-J5/3

1580; [*Valtellina*]; *Venice, Francesco Valeggio; n.i.* Tooley, *NL 4F318.*

110. Fleur-de-lys in circle under six-pointed star D; circle 39 x 42 mm; total height 70 mm. G6/2[U1]-J5/3

1585; [*Venice*]; *Venice, Donato Rascicotti; n.i.* Tooley, *NL 4F292.*

111. Fleur-de-lys in two concentric circles under shield; circles inner dia: 41 mm, outer dia: 51 mm; total height 67 mm. G6/2[2*(U1)]M13

1613; *[Spoleto]; Rome, Filippo Thomassino; n.i.* *TOOLEY, NL 4F279.*

112. Fleur-de-lys over triple-mount in circle with single-wire 'M' over circle A; circle 44 x 45 mm; total height 56 mm. G6/2J3/3[U1{c:"M"}]. Other manuals: HEAWOOD CAT 1620 (sim.)

1580; *[Trapani]; Rome, Henricus van Schoel; n.i.* *TOOLEY, NL 4F282.*

111

112

113

114

113. Fleur-de-lys over triple-mount in circle with single-wire 'M' over circle B; circle 44 x 46 mm; total height 56 mm. G6/2-J3/3[U1{c:"M"}]

n.d. *[Orvieto]; n.pl., Giovanni Battista de Rossi S.J; n.i.* TOOLEY, *NL 4F236.*

114. Fleur-de-lys in shield; shield height 83 mm; with crossed keys. G6/2-[2*(x:N16)]<f:23>. Other manuals: ZONGHI 590

1566; *[Malta]; Rome, Antonio Lafreri;* TOOLEY *361, NL 2F183.*

115. Fleur-de-lys in shield with six-pointed star A; shield height 84 mm; with crossed keys; total height 105 mm. G6/2[(x:N16<f:M13{b:J5/4}>]. Other manuals: Beans 52 (sim.); Heawood Cat 2824 (sim.)

__1582__; [Acquapendente]; Rome, Paulo Gratiano; Tooley 95, NL 2F113.

116. Fleur-de-lys in shield with six-pointed star B; shield height 79 mm; with crossed keys; total height 100 mm. G6/2[(x:N16<f:M13{b:J5/4}>]. Other manuals: Heawood Cat 2824 (sim.)

__1580__ ca; [Piacenza]; Rome, Giovanni Orlandi; n.i. Tooley, NL 4F243.

115

116

117. Fleur-de-lys in shield with six-pointed star C; shield height 80 mm; with crossed keys; total height 102 mm. G6/2[(x:N16<f:M13{b:J5/4}>]. Other manuals: Heawood Cat 2824 (sim.)

***1590** ca; [Western Asia]; n.pl., Giacomo Gastaldi (Antonio Lafreri, Petri de Nobilibus, Paulo Gratiano); Tooley 51, GHB 7043r.*

118. Fleur-de-lys in shield with six-pointed star D; variant of no. 117. G6/2[(x:N16<f:M13{b:J5/4}>]

***1590** ca; [Western Asia]; n.pl., Giacomo Gastaldi (Antonio Lafreri, Petri de Nobilibus, Paolo Gratiano); Tooley 51, Boswell 7043v*

117

118

119. Tulips in circle with four-pointed star A; three tulips (each with three petals) on one stem; circle 45 x 46 mm; star 7 mm above; total height 78.5 mm. G9[<f:U1{b:J5/2}>]

1542; [*Perpignan*]; *n.pl.*, [*Enea Vico*]; *Tooley 445, NL 2F26.*

120. Tulips in circle with four-pointed star B; three tulips (each with three petals) on one stem; crooked circle 45 x 47 mm; star 9 mm above; total height 82 mm. G9[<f:U1{b:J5/2}>]

1542; *Perpignian Carcasona Nerbona*; *n.pl.*, *E*[*nea*] *V*[*ico*]; *Tooley 445,* NC *2:130:33*

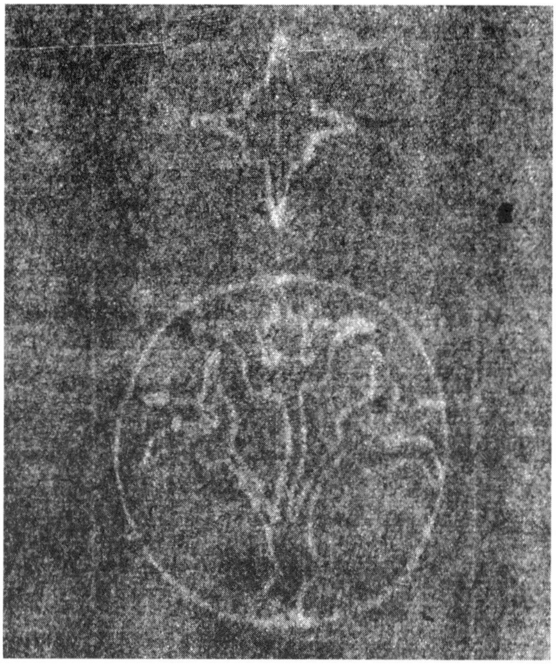

119

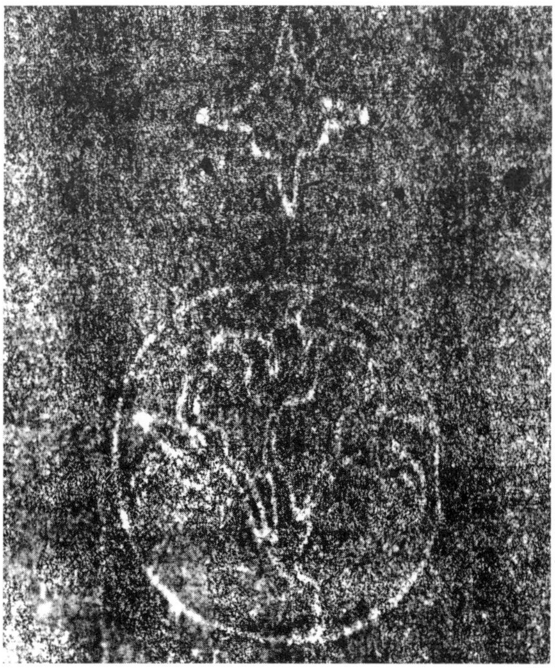

120

121

122

121. Tulips in circle with six-pointed star A; three tulips (each with three petals) on one stem; circle 46 x 48 mm; total height 75 mm. G9[<f:U1{b:J5/4}>]

__1558__; [Italy]; Rome, Vincenzo Lucchino; Tooley 327, NL 4F199r.

122. Tulips in circle with six-pointed star B; three tulips (each with three petals) on one stem; circle 48 mm dia; total height 75 mm; twin of #123? G9[<f:U1{b:J5/4}>]. Other manuals: Beans 19

1558; [Italy]; Rome, Vincenzo Lucchino; Tooley 327, NL 4F199v.

__1558__; [Greece]; Rome, Vincenzo Lucchino; Tooley 279, NL 4F355.

123. Tulips in circle with six-pointed star C; three tulips (each with three petals) on one stem; circle 49 mm dia; total height 75 mm. G9[<f:U1{b:J5/4}>]. Ilkley print.

1566; [Austria and the Balkans]; Venice, Paolo Forlani; Tooley 144, Doria Atlas 45.

124. Tulips in shield under six-pointed star A; three tulips (each with three petals) on one stem; shield 39 x 78 mm; total height 107 mm. NB. central wire passes through higher end of slanting base of stem; rightmost petal touches shield side. G9[<f:M13{b:J5/4}>]. Other manuals: Beans 37 (sim.)

1542; [Budapest]; n.pl., Enea Vico; Tooley 150, GHB 209.

1557; *[Rome]; Rome, Antonio Lafreri; Tooley 472, NL 2F130.*

1560; [Rome]; n.pl., n.pr.; Tooley 475, NL 2F166.

1564; [Lombardy]; Rome, Antonio Lafreri; Tooley 352, GHB 221.

1565; [Malta]; Rome, Antonio Lafreri; Tooley 360, NC 2:130:73.

1565; [Malta]; Rome, Antonio Lafreri; Tooley 383, CSUF ac1334.

1570; [Eastern Mediterranean]; Rome, Antonio Lafreri; Tooley 184, GHB 223r.

1570; [Lombardy]; Rome, Antonio Lafreri (Giacomo Gastaldi); Tooley 354, NL 4F309.

1570 ca; [Cyprus]; n. pl., n. pr.; Tooley 181, GHB 227

n.d. [Austria and Hungary]; n.pl., n.pr.; Tooley 121, GHB 208.

n.d. [Algiers]; n.pl., n.pr.; Tooley 97, GHB 248.

n.d. [Boulogne]; n.pl., n.pr.; Tooley 140, NL 2F27.

n.d. [Algiers]; n.pl., n.pr.; Tooley 97, NL 2F238.

n.d. [Frisland]; Rome, Antonio Lafreri (orig.?); Tooley 222, NC 2:130:[map 6 left].

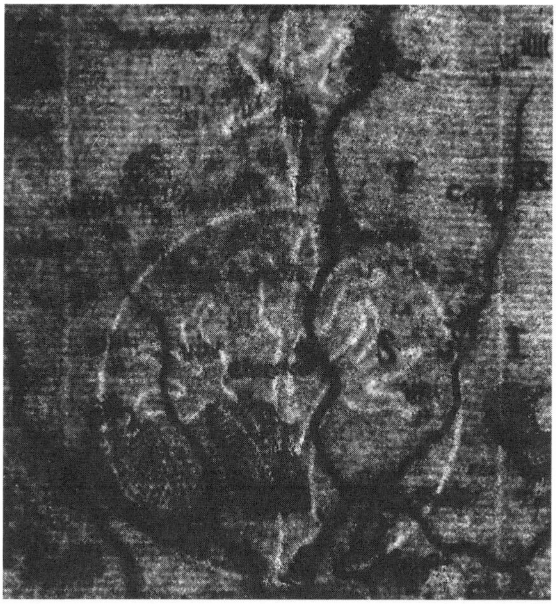

123

124

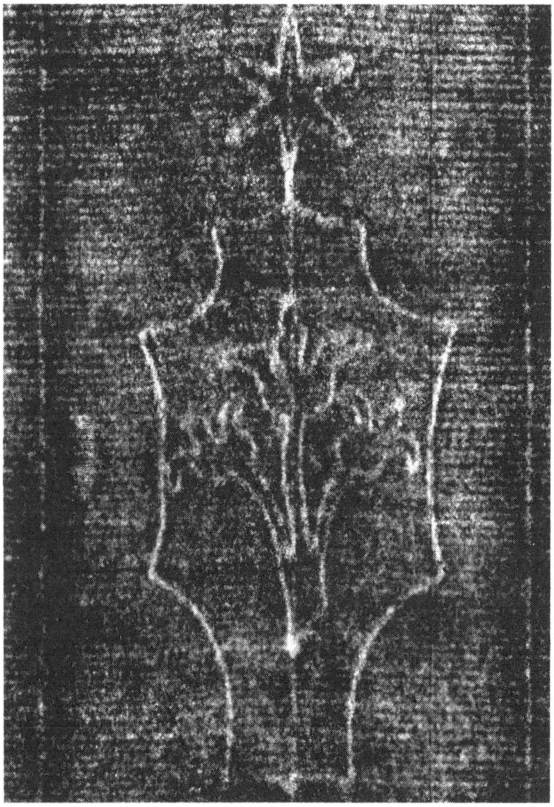

125

125. Tulips in shield under six-pointed star B; three tulips on one stem (each with three petals); shield 43 x 77 mm; total height 102 mm. NB. central wire passes through the lower end of slanting base of stem. G9[<f:M13{b:J5/4}>]

1557; [Nettuno]; n.pl., n.pr.; Tooley 417, NL 2F170.

1558; [Guines]; Rome, n.pr.; Tooley 293, GHB 237.

1559; [Hungary]; Rome, Antonio Lafreri (reduced from 1556 Lazius map); Tooley 307, NC 2:130:28 (ms).

1560; [Rome]; Rome, Antonio Lafreri (orig.?); Tooley 475, NC 2:130:40.

1564; [Hispaniola]; Venice, Paolo Forlani; Tooley 84, NL 4F83.

1564; [Ancona]; Rome, Vincenzo Lucchino; Tooley 102, NL 2F165.

1565; [Malta]; Rome, Antonio Lafreri; Tooley 384, NL 2F41.

n.d. *[Rhodes]; n.pl., n.pr.; Tooley 463, NL 2F201.*

n.d. [Greece, etc.]; n.pl., n.pr.; Tooley 290, NL 2F22.

n.d. [Ireland]; n.pl., n.pr.; Tooley 315, NL 2F72.

n.d. [Ireland]; Rome, Claudio Duchetti; Tooley 318, NL 2F74.

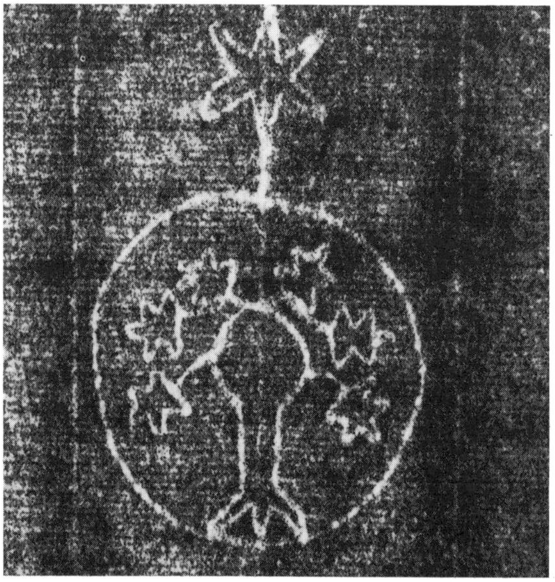

126

126. Tree in circle under six-pointed star A; tree with six leaves in crooked circle 46 mm dia; total height 71 mm. H3[(6*u05<f:U1{b:J5/4}>)]

1565*; [Malta]; Venice, Domenico Zenoi; Tooley 387, NL 2F46.*

127. Tree in circle under six-pointed star B; tree (with thin trunk) with six leaves; crooked circle 44 x 45 mm under star; total height 72 mm. H3[(6*u-05<f:U1{b:J5/4}>)]

1565; *[Malta]; Venice, Domenico Zenoi; TOOLEY 387, NL 2F45.*

128. Tree in circle under six-pointed star C; tree with six leaves (lower branches crossing); circle 47 x 48 mm; total height 73 mm. H3[(6*u05<f:U1{b:J5/4}>)]. Other manuals: BEANS 8.

1565; [World]; Venice, Paolo Forlani; TOOLEY 11, NL 4F10r.

1565; [World]; Venice, Paolo Forlani; TOOLEY 11, NL 4F10v.

[***1565***]; *[North America – New France]; Venice, Bolognino Zaltieri [Paolo Forlani]; TOOLEY 81, NL 4F56.*

[**1565**]; [North America – New France]; Venice, Bolognino Zaltieri [Paolo Forlani]; Tooley 81, NMC 22901.

n.d. [Tunis]; n.pl., n.pr.; TOOLEY 559, NL VCA 31.

Terminus ad quem is likely to be 1565.

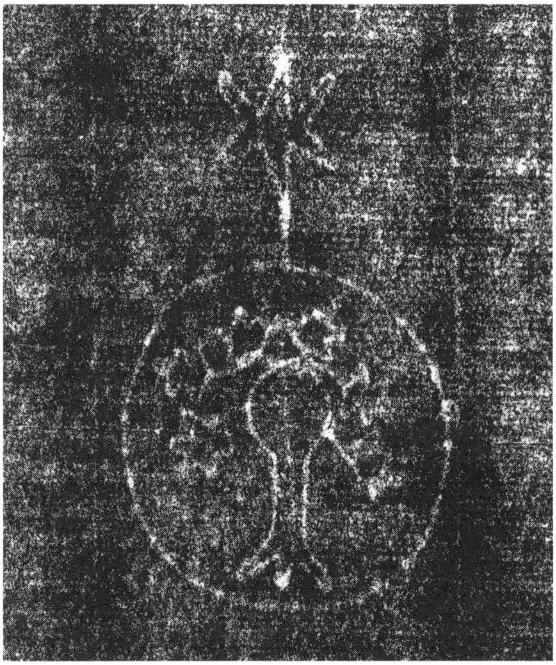

127

128

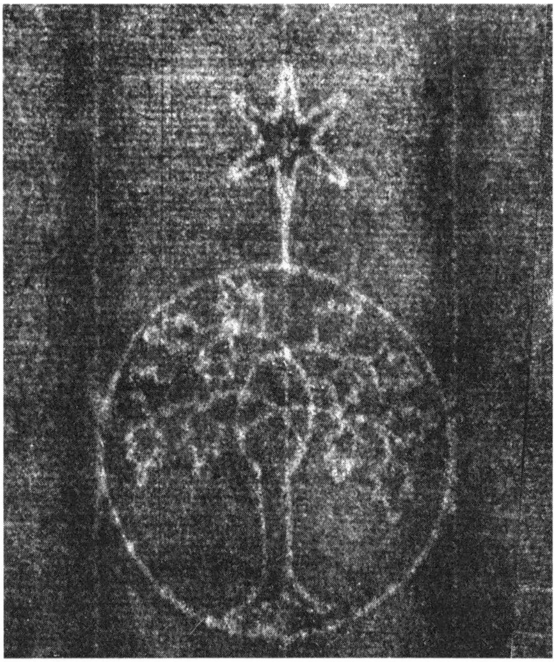

129

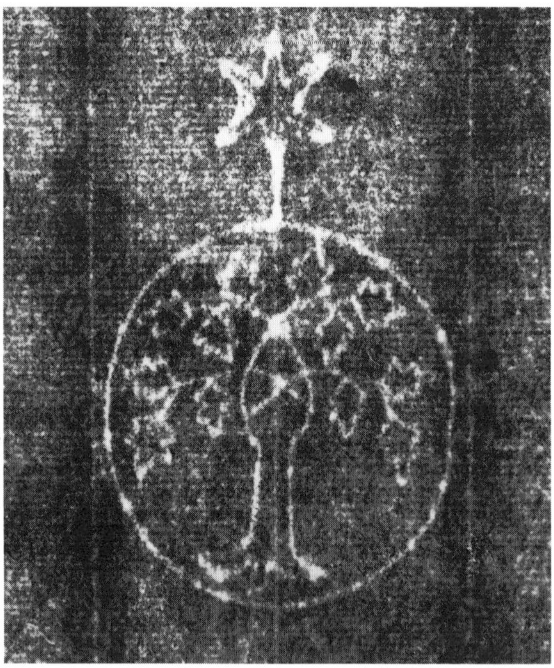

130

129. Tree in circle under six-pointed star D; twelve leaves, lower branches crossing; circle 48 x 49 mm; total height 75mm. H3[(12*u05<f:U1{b:J5/4}>)]. Other manuals: HEAWOOD CAT 3955 (sim.)

1562; *[Africa] right sheet; Northwestern University (internal watermark number 30779/2)*

130. Tree in circle under six-pointed star E; similar to #129, but circle not touching chain lines. H3[(12*u-05<f:U1{b:J5/4}>)]. Other manuals: BEANS 8 (sim.); HEAWOOD CAT 3955 (sim.)

1560; [Africa]; [Venice], Paolo Forlani; TOOLEY 67, NL 4F392.

1562; *[Africa] left sheet; Northwestern University (internal watermark number 070379/1)*

129 and 130 are likely twins.

131. Tree in Shield A; tree above mountains in curved-sided shield (42 x 75mm). H3[{b:J3/3}]<f:M13>

1564; [Greece, 2 sheets]; Venice, Ferrando Bertelli (Domenico Zenoi); Tooley 286, Doria Atlas 91.
1565; [Southeast Asia]; [Venice], Ferrando Bertelli; Tooley 73,
GHB 121(iii).
1565; *[Africa]; [Venice], Ferrando Bertelli; Tooley 73; GHB 121(i).*
n.d. [Atlantic Ocean]; [Venice], Ferrando Bertelli; Tooley 78, Doria Atlas 6.

132. Tree in Shield B; tree above mountains in straight-sided shield (47 x 75mm) with pointed shoulders, one of which is bent almost horizontal. H3[{b:J3/3}]<f:M13>. Other manuals: Beans 30 (probably). This is probably the watermark after which George H. Beans named his "Tall Tree Library."

1564; [Corfu]; n.pl., Ferrando Bertelli; Tooley 165, NL 2F188.
1564; [Greece, 2 sheets]; Venice, Ferrando Bertelli, Tooley 286;
Doria Atlas 91(ii).
1565; [Southeast Asia]; n.pl., Ferrando Bertelli; Tooley 73, GHB 121(ii).
1565; [Valletta]; Venice, Niccolo Nelli; n.i. Tooley, GHB 400.
n.d. *[Atlantic Ocean]; n.pl., Ferrando Bertelli, Tooley 78; NL 2F246.*

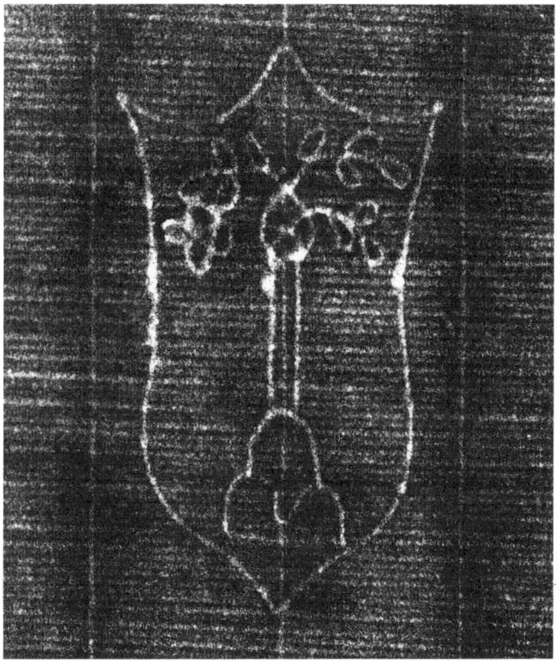

131

132

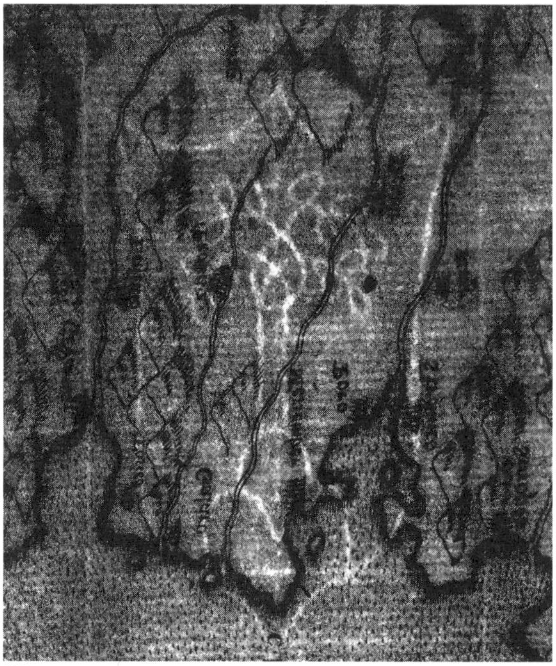

133

134

133. Tree in Shield C; tree above mountains in straight-sided shield (46 x 75 mm) with pointed shoulders. H3 [{b:J3/3}]<f:M13>. Ilkley print.

1544; [Spain]; Venice, Giacomo Gastaldi; Tooley 527, Doria Atlas 18.

1562; [British Isles]; Venice, Ferrando Bertelli; Tooley 272, Doria Atlas 15.

1566; [North America – New France]; Venice, Bolognino Zaltieri, Tooley 80; Doria Atlas 108.

1566; [North America – New France]; Venice, Bolognino Zaltieri, Tooley 80; NMC 22900.

***1567**; [Austria and Balkans]; Venice, Domenico Zenoi; Tooley 125, Doria Atlas 43g.*

n.d. [Sicily]; Venice, Ferrando Bertelli; Tooley 521, Doria Atlas 84.

134. Tree in Shield D; tree above mountains in curved-sided shield (48 x 69mm) with pointed shoulders; central tip not on chain wire. H3[{b:J3/3}]<f:M13>

1545; [Austria and Upper Rhine]; Vienna, (Lazius); n.i. Tooley, Doria Atlas 50.

1545; [Sicily]; Venice, Giacomo Gastaldi; Tooley 514, Doria Atlas 83.

1563; [Tuscany]; Venice, Ferrando Bertelli; Tooley 574, Doria Atlas 68.

***1566**; [Tunis]; Venice, Bolognino Zaltieri; Tooley 558, NL 2F243.*

1566; [France, 2 sheets]; Venice, Bolognino Zaltieri; Tooley 211,
Doria Atlas 23(ii).

1567; [Brabant/Belgium]; Venice, Bolognino Zaltieri; Tooley 145, Doria Atlas 28.

135. Tree in Shield E; tree with 14 leaves, crossed branches, above mountains in simple shield (38 x 66 mm). H3[{b:J3/3}]<f:M13>

1560; [Tuscany, Appennines, Sicily]; Rome, Antonio Lafreri (Venice/Giacomo Gastaldi orig.); TOOLEY 28, SW sheet, NC 2:130:42 left (bottom only).

1560; [Tuscany, Appenines, Sicily]; Rome, Antonio Lafreri (Venice/Giacomo Gastaldi orig.); TOOLEY 28, SW sheet, NC 2:130:42 right.

1560; [Macedonia, part of Asia]; Rome, Antonio Lafreri (Venice/Giacomo Gastaldi orig.); TOOLEY 28, SE sheet, NC 2:130:51 left (top only).

1560; *[Macedonia, part of Asia]; Rome, Antonio Lafreri (Venice/Giacomo Gastaldi orig.); TOOLEY 28, SE sheet, NC 2:130:51 right.*

136. Bunch of grapes, with letters; triangular bunch, with 1-2-3-4-5-6-7-6-5-4 grapes in each row; curved stem; 'A' to left, 'R' to right, both in double outline; 30 x 52mm. H4/2. Other manuals: BRIQUET 13166

1561; [Portugal]; Rome, Michele Tramezzino; TOOLEY 456, NL 4F195ms.

n.d. *[Transylvania and Hungary]; n.pl., Domenico Zenoi (Giovanni Francesco Camocio); TOOLEY 553, GHB 251v.*

135

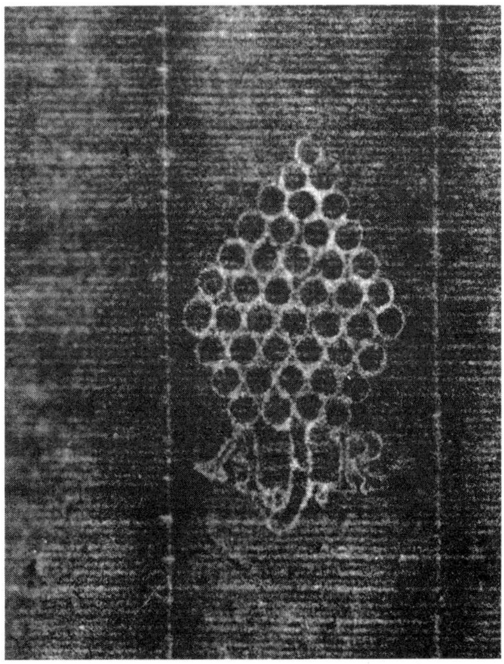

136

137

137. Two crescents, one (29mm high) held within the horns of the other (71mm high). (2*(J2/1))

1599; [*Italy*]; *n.pl., Giacomo Gastaldi; n.i.* TOOLEY, *NL 2F106v.*

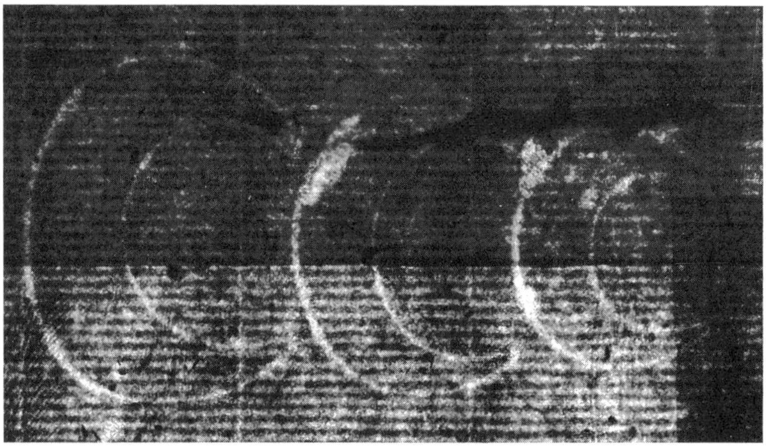

138

138. Three crescents in a row A; shrinking in size; respective heights: 45.5, 40, 32mm; set between four chain lines; one horn of smallest crossing the fourth wire. (3*(J2/1))

1567; [*Messina*]; *Rome, Antonio Lafreri;* TOOLEY *388, NL 4F228(3).*

139. Three crescents in a row B; Similar to # 144, but thinner crescents; heights: 43, 39, 32 mm. (3*(J2/1))

1567; [Messina]; Rome, Antonio Lafreri; Tooley 388, NL 4F228(2).

1567; *[Messina]; Rome, Antonio Lafreri; Tooley 388, NL 4F228(4).*

1567; [Messina]; Rome, Antonio Lafreri; Tooley 388, NL 4F228(5).

140. Three crescents in a row C; wide crescents in a row; largest touching a chain line, but smallest does not reach fourth wire; heights: 38, 30, 27mm. (3*(J2/1))

n.d. *[World]; n.pl./printer; n.i. Tooley, NL 4F16(1 bottom).*

139

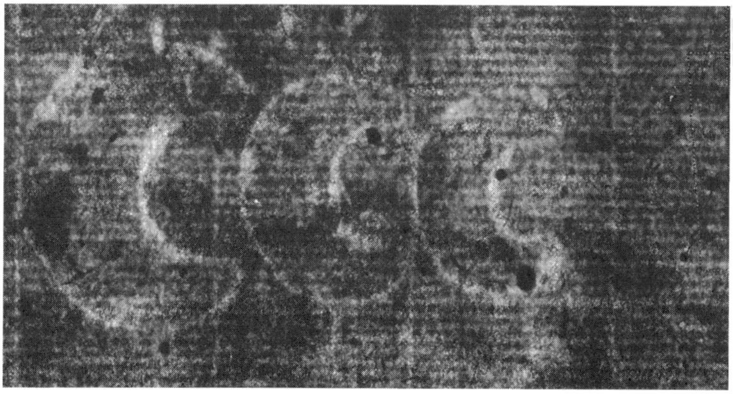

140

141. Three crescents in a row D; similar to #140 but smaller; heights: 36, 31.5, 21(?) mm. (3*(J2/1))

n.d. *[World]; n.pl., n.pr.; n.i.* TOOLEY, *NL 4F16(2 bottom).*

142. Three crescents in a row E; thinner crescents in a row; decreasing in height (34, 34, 25 mm). (3*(J2/1))

1597*; [Brescia, 5 sheets]; n.pl., Leone Pallavicini; n.i.* TOOLEY, *NL 4F299.*

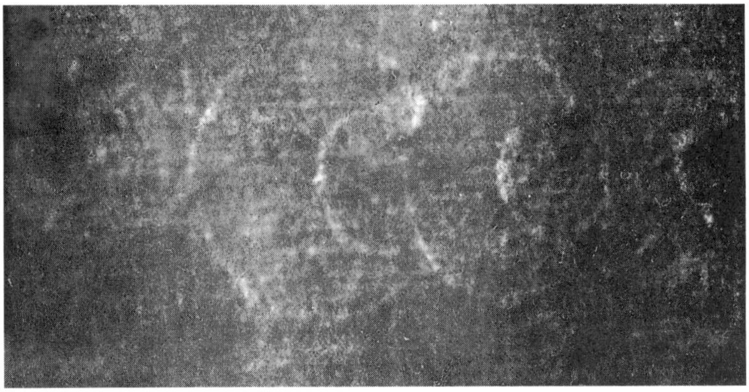

141

142

143. Three crescents in a row F; large crescents in a row; decreasing in height (56, 46, 37 mm); middle crescent has flat "back". (3*(J2/1))

1574; [*Venice*]; *n.pl., Ferrando Bertelli; n.i.* *Tooley, NL 4F291.*

144. Three crescents in a row G; crescents in a row; decreasing in height (39, 31, 26 mm). (3*(J2/1))

1560 *ca;* [*Lombardy*]; *Venice, Giacomo Gastaldi (Stefano Scolari); n.i.* *Tooley, NL 4F304.*

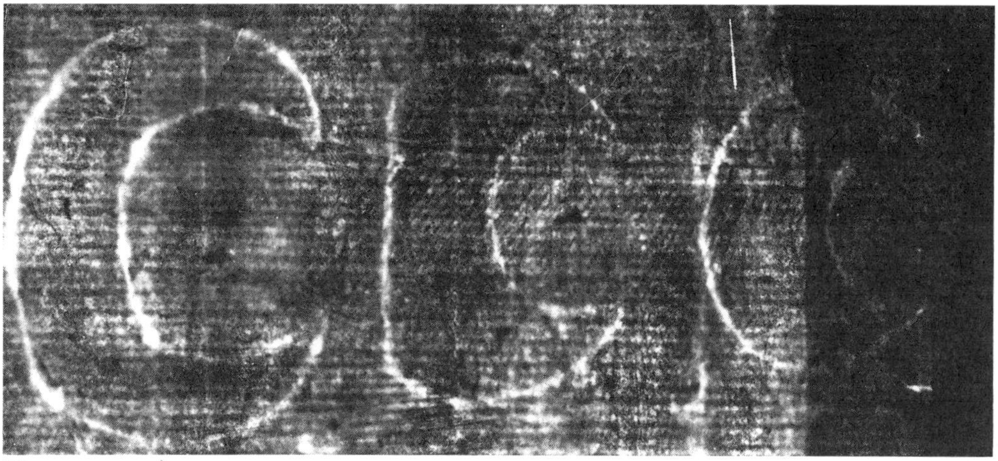

143

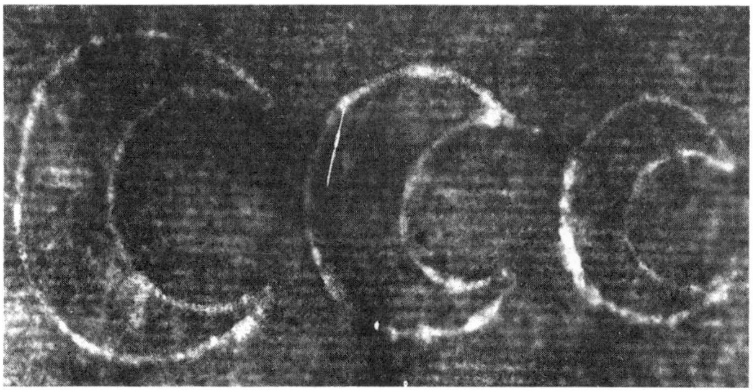

144

145

146

145. Five unevenly spaced crescents (horns pointing upwards); in cross in a crooked oval 46 x 50 mm; under fleur-de-lys; total height 85 mm. (5*(J2/1)). Other manuals: BEANS 10; HEAWOOD CAT 883; BRIQUET 5380 (sim.)

1560; *[Spain]; Venice, Domenico Zenoi; TOOLEY 532. NL 4F190.*

1560; [World]; Rome, Antonio Lafreri; TOOLEY 2. NL 4F7.

[**1565**]; marginal strips and mounting paper used for composite atlas in the Biblioteca Nazionale Marciana, Venice (138.c.4); datable through other internal evidence as 1565.

146. Three mountains in circle under six-pointed star; circle 44 mm dia. J3/3<U1>-{t:J5/4}. Other manuals: HEAWOOD CAT 2610 (sim.); BRIQUET 11932 (sim.)

1575; *[Rome (Piazza San Pietro)]; Rome, Giovanni Battista de Cavalleri; n.i. TOOLEY, NL 4F270.*

147. Six mountains in pyramidal formation; height 25mm. J3/4

__1590__ ca; [Venice]; n.pl., Matteo Florimi; SCHULZ *55,* VAN DER HEYDEN *47, NL 4F286.*

148. Six-pointed star A; six long points around hub (12mm dia); filling circle (50 x 51 mm) under simple cross; total height 75 mm. J5/4<f:U1{t:S1}>. Other manuals: BEANS 53 (sim.); BRIQUET 6089 (sim.)

__1592__ ca; [Title Page]; Rome, Antonio Lafreri (Petri de Nobilibus); TOOLEY *[unnumbered first entry], Boswell 175r.*

147

148

149

150

149. Six-pointed star B; long points; filling circle, 49 x 51 mm. J5/4<f:U1>

n.d. Tabula moderna Poloniae, [Rome], Antonio Salamanca; n.i. TOOLEY, NC 1:48:15.

1557; *Illustriss. Atq. Integerrimo Octavio farnesio; Rome, Antonio Lafreri; n.i.* TOOLEY, *NC 1:48:47.*

150. Six-pointed star C; long points; filling circle, 50 x 48 mm, under star; height 78 mm. J5/4<f:U1{t:J5/4}>

n.d. *[Plan of unidentified fortress] 440 x 304 mm; n.pl., n.pr.,* TOOLEY *315,* NC *1:48:7.*

151. Six-pointed star D; in lozenge inside circle; four smaller six-pointed stars between edge of square and circle. .J5/4<f:U3<f:U1-(4*(J5/4))>>.
Other manuals: BEANS 6 (possible); BRIQUET 6098 (sim.), BORRONI SALVADORI 90 (sim.)

1553; [Piedmont]; n.pl., n.pr.; TOOLEY 447, NL 4F298.

152. Sun in circle with wavy rays in double wire; face in central circle. J6/2/2<f:U1>

1595 ca; [Perugia]; n.pl., Valerio Spada; n.i. TOOLEY, NL 4F326.

151

152

153

154

153. Castle on triangular base A; 29 x 36 mm. K7<b:U5>. Other manuals: BRIQUET 15958; PICCARD III:III:231, 232 (sim.)

__1567__; Gotha; n.pl., n.pr.; TOOLEY 263, NC 1:48:36 left.

154. Castle on triangular base B; 52 x 65 mm. K7<b:U5>. Other manuals: BRIQUET 15931 (sim.)

1565; [*Malta*]; *Nuremberg, Mathias Zündten; n.i. TOOLEY, NC 1:48:24 right*

155. Castle on triangular base C; 52 x 71 mm. K7<b:U5>. Other manuals: BRIQUET 15931 (sim.)

1565; *[Malta]; Nuremberg, Mathias Zündten; n.i. TOOLEY, NC 1:48:24 left*

156. Anchor in double outline A; unencircled. L4/4/1/2. Other manuals: MOSIN type 1, 336

n.d. [Goletta]; n.pl., n.pr.; TOOLEY 261, GHB 246.

n.d. *Golletta; n.pl., n.pr.; n.i. TOOLEY, NL 2F241.*

155

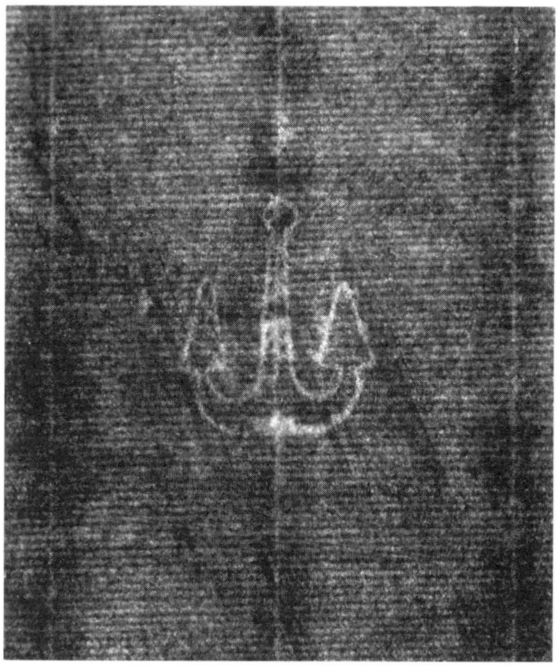

156

157

158

157. Anchor in double outline B; in circle. L4/4/1/2<f:U1{t:J5/4}>. Other manuals: Beans 22; Mosin type 1, 430 (sim.)

n.d. *[Transylvania and Hungary]; n.pl., Domenico Zenoi (Giovanni Francesco Camocio); Tooley 553, GHB 251r.*

158. Anchor in double outline with star A; 45 mm dia circle; 30 mm between chain lines. L4/4/1/2<f:U1{t:J5/4}>. Other manuals: Mosin Type It. II.2A.d, Beans 23.

1561; [Italy]; Venice, Giacomo Gastaldi; Tooley 328, NL 4F200r.

1566; *[Naples]; Rome, Antonio Lafreri; Tooley 408, CSUF 1326.*

n.d. [Thionville]; n.pl., n.pr.; Tooley 548, NL 2F30.

159. Anchor in double outline with star B; 45 mm dia circle; 24 mm between chain lines; central chain line bent. L4/4/1/2<f:U1{t:J5/4}>

1564; *[Asia Minor]; Venice, Giacomo Gastaldi; Tooley 64a, GHB 157r.*

1565; [Ancona]; Venice, Ferrando Bertelli; Tooley 104, Doria Atlas 66.

1566(?); [Siege of Szigetvár]; n.pl., n.pr.; n.i. Tooley; Ganado 1993, 20; Doria Atlas 46.

160. Anchor in double outline with star C; 24 mm between chain lines; central line straight. L4/4/1/2<f:U1{t:J5/4}>. Other manuals: Beans 23

1563; [Holland]; Venice, Paolo Forlani; Tooley 301, GHB 121.

1563; [Rome]; Venice, Ferrando Bertelli (Paolo Forlani); Tooley 480, NL 4F327.

1564; [Asia Minor]; Venice, Giacomo Gastaldi; Tooley 64a,

GHB 157r.

1565; *[Ireland]; Venice, Bolognino Zaltieri; Tooley 316, NL 2F73.*

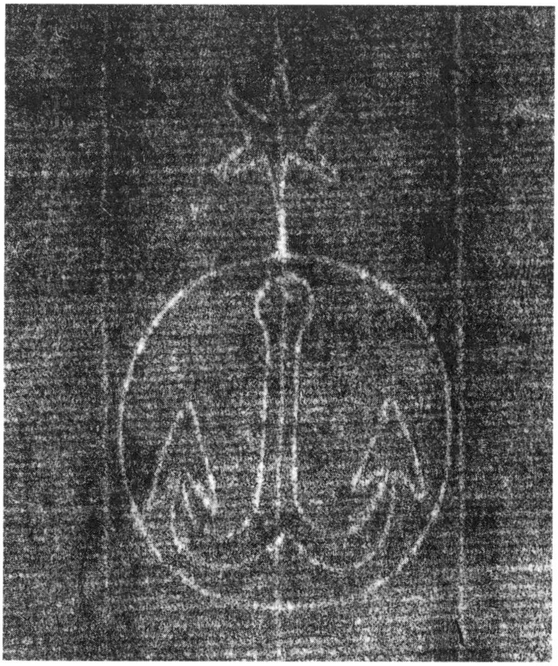

159

160

161

161. Anchor in double outline with star D; in circle (35 mm dia); star above. L4/4/1/2<f:U1{t:J5/4}>

1561; *[Piedmont]; n.pl., Giacomo Gastaldi; Tooley 449, NL 4F314.*

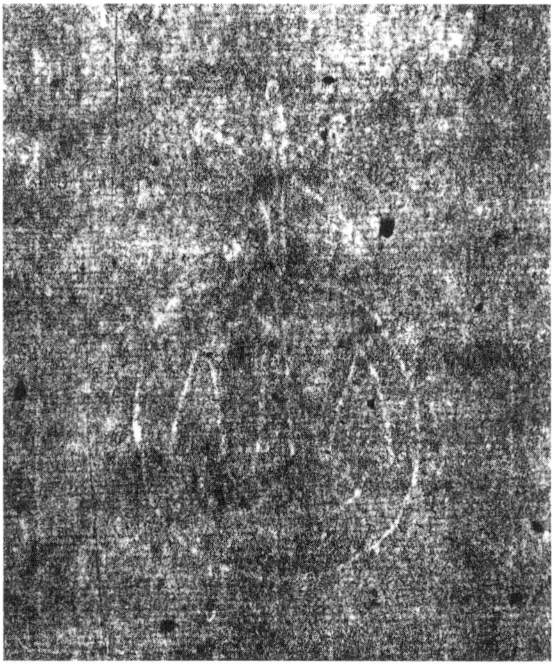

162

162. Anchor in double outline with star E; in circle (38 mm dia); star above. L4/4/1/2<f:U1{t:J5/4}>

n.d. *[Constantinople]; n.pl., Giovanni Andrea Vavassore; n.i. Tooley, Boswell 378.*

163. Anchor in double outline with star F; in circle (40 mm dia); star above. L4/4/1/2<f:U1{t:J5/4}>

1571; *IL vero ordine dell'armata Christiana e Turchesca Ordinata il giorno della battaglia che fu addì 7 ottobre 1571; n.pl., n.pr.; n.i.* TOOLEY, *NL 2F61.*

164. Anchor in double outline with star G; in circle (43 mm dia); star above. L4/4/1/2<f:U1{t:J5/4}>

n.d. *[Austria and Hungary]; n.pl., n.pr.;* TOOLEY *121, NL 4F175.*

1562; [World]; n.pl., n.pr.; TOOLEY 8, NL 4F8.

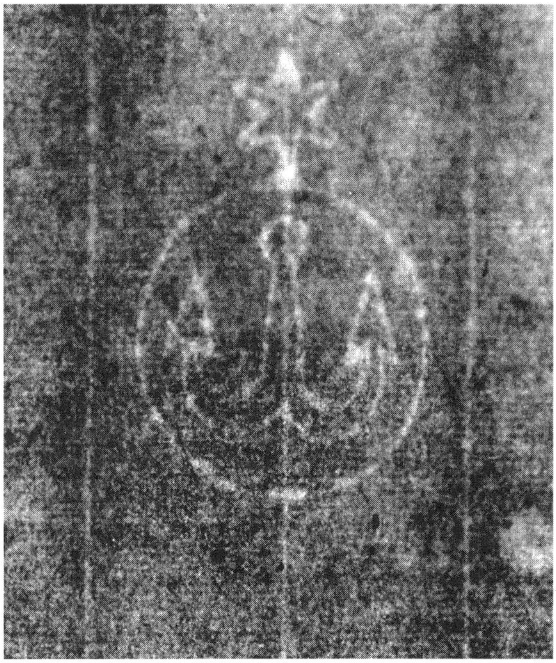

163

164

165

166

165. Anchor in double outline with star H; in oval (37 x 29 mm dia); star above. L4/4/1/2<f:U1{t:J5/4}>

1554; *[South Asia]; n.pl., Giacomo Gastaldi (Giovanni Battista Ramusio); n.i. Tooley, KN 002-RN-6r.*

166. Anchor in double outline with star I; in oval (damaged, 40 x 30 mm dia); star above. L4/4/1/2<f:U1{t:J5/4}>

1562; *[Russia]; Venice, Giacomo Gastaldi (Ferrando Bertelli); Tooley 499, NL 2F215.*

167. Anchor in double outline with star J; in oval (40 x 46 mm dia); star above. L4/4/1/2<f:U1{t:J5/4}>

1543; [*Nice*]; *n.pl., Enea Vico; Tooley 419, NL 4F95.*

168. Anchor in double outline with star K; in circle (36 x 35 mm); star above; total height 61mm. L4/4/1/2<f:U1{t:J5/4}>. Other manuals: Briquet 485 (sim.)

1557; [*Friuli*]; *Venice, Giovanni Andrea Vavassore; n.i. Tooley,* NC *1:48:19.*

167

168

169

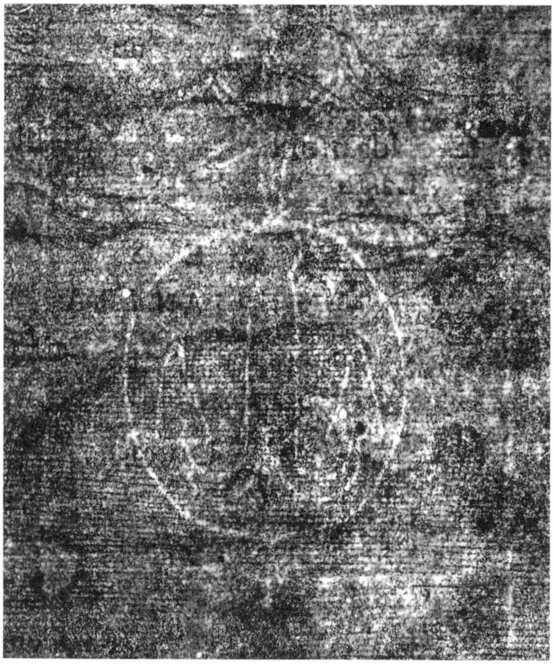

170

169. Anchor in double outline with star L; in circle (39 x 33 mm, crooked); height 57mm. L4/4/1/2<f:U1{t:J5/4}>

1568; [*Poland, second part*]; *Venice, Giacomo Gastaldi (Paolo Forlani)*; *Tooley 455*, NC *2:130:23*.

170. Anchor in double outline with star M; in circle (38 x 41 mm); total height 64mm. L4/4/1/2<f:U1{t:J5/4}>

1538; [*Crete, Lisola de Candia*]; *Venice, Giovanni Andrea Vavassore (Matteo Pagano)*; *Tooley 172*, NC *1:48:27*.

171. Anchor in double outline with star N; in circle (37 x 36 mm); total height 58 mm. L4/4/1/2<f:U1{t:J5/4}>. Other manuals: BRIQUET 486 (sim.); MASSARI (1983) 105b (sim.)

1538 *Isola di Cipro; Venice, Matteo Pagano; TOOLEY 179,* NC *1:48:26 right.*

172. Anchor in double outline with star O; in circle (38 x 40 mm); total height 60 mm. L4/4/1/2<f:U1{t:J5/4}>

1622; [*Tivoli*]; *Rome, Antonius Paolis; n.i. TOOLEY, NL 4F280.*

171

172

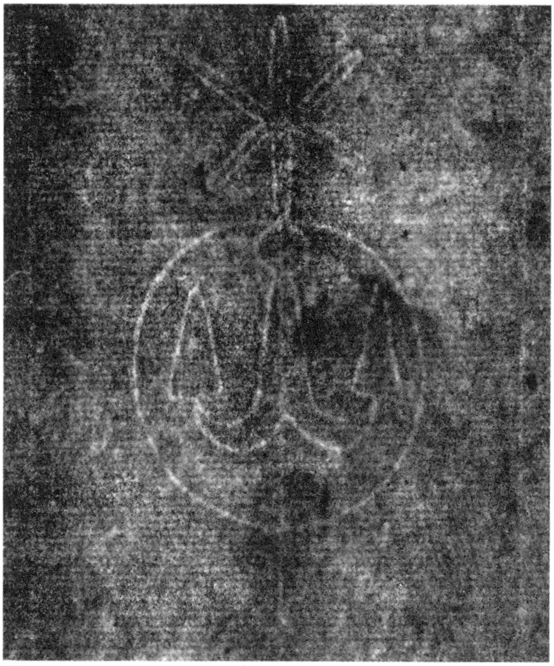

173

174

173. Anchor in double outline with star P; slightly off-center in circle (39 x 41 mm); star above; total height 68mm. L4/4/1/2<f:U1{t:J5/4}>. Other manuals: Briquet 485 (sim.)

1545; [Sicily]; Venice, Giacomo Gastaldi; Tooley 514, NL 4F347.

1545; *[Sicily]; Venice, Giacomo Gastaldi; Tooley 514, NL 4F348.*

174. Anchor with double-lined arms A; single-line shank, in circle (41x37 mm); star above. L4/4/1/2<f:U1{t:J5/4}>. Other manuals: Mosin Type It.III.a; Briquet 548 (sim.); Beans 57 and 66 (somewhat sim.)

1567; *[Malta]; n.pl., Domenico Zenoi; Tooley 358, GHB 278.*

175. Anchor with double-lined arms B; single-line shank, in circle; star above; 'M' below [MOSIN Type It.III.a]. L4/4/1/2<f:U1{t:J5/4}>

1567; [*Parma*]; *Venice, Paolo Forlani;* TOOLEY *443, NL 2F49.*

176. Anchor with double-lined arms C; single-line shank, in circle (44 x 40mm); star above; total height 66mm. L4/4/1/2<f:U1{t:J5/4}>

1562; [*Africa*]; *Venice, Paolo Forlani;* TOOLEY *67,* NC *2:130:53 left.*

175

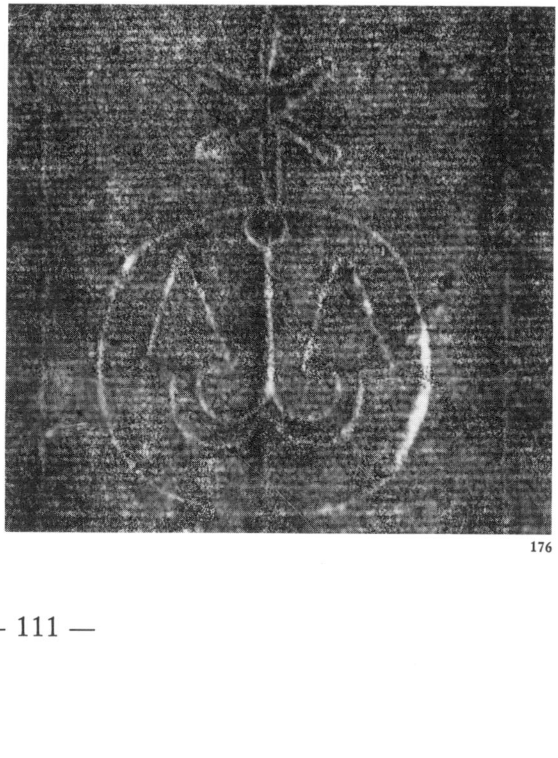

176

177

178

177. Anchor with single-lined arms, shank, in circle A; circular trefoil. L4/4/1/2<f:U1{t:G4}>. Other manuals: MOSIN Type IV 1983 (sim.)

1567; [Mirandola]; Venice, Paolo Forlani; TOOLEY 396; NL 2F121r.

***1600** ca); Reggio di Lombardia; n.pl., Justus Sadeler; n.i. TOOLEY, NL 2F128.*

178. Anchor with single-lined arms, shank, in circle B; lozenge trefoil. L4/4/1/2 <f:U1{t:(3*(U3))}>. Other manuals: MOSIN Type IV 2307 (sim.)

***1580**; [Ancient Rome]; Venice, Onofrio Panvinio; n.i. TOOLEY, KN 051-21-#8.*

1594; Il vero retratto della fortezza di Vienna; n.pl., n.pr.; n.i. TOOLEY, NL 2F94.

179. Anchor with single-lined arms and shank in circle with lozenge quatrefoil A; small double-sewing dot at foot. L4/4/1/2 <f:U1{t:(4*(U3))}>. Other manuals: Beans 81 (sim.), Mosin Type It.IV.2.i 2518 (sim.)

1566; [Switzerland]; Venice, Paolo Forlani; n.i. Tooley, Boswell 179v.

1566; [Dalmatia]; n.pl., n.pr.; Tooley 191, Boswell 184r.

ca. 1585; [Friuli]; Venice, Donatum Bertelli; Tooley 226, Boswell BHS 193r.

***1585** ca; [Istria]; Venice, n.pr.; Tooley 325, Boswell 206r*

180. Anchor with single-lined arms and shank in circle with lozenge quatrefoil B; long sewing patterns on both sides. L4/4/1/2<f:U1{t:(4*(U3))}>. Other manuals: Beans 81 (sim.), Mosin Type It.IV.2.i 2518 (sim.)

***1566**; [Switzerland]; Venice, Paolo Forlani; n.i. Tooley, Boswell 180r.*

179

180

181

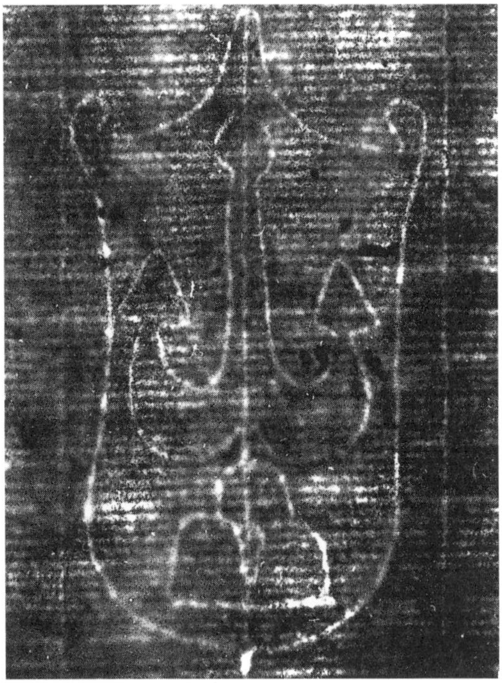

182

181. Anchor with double outline, in shield A; 44 x 91 mm; triple-mount below. L4/4/1/2{b:J3/3<f:M13>}. Other manuals: Mosin Type VI 2636 (sim, but without star)

1592 *ca; [Tripoli]; Rome, Claudio Duchetti; Tooley 554, GHB 201*

182. Anchor with double outline, in shield B; 48 x 88 mm; triple-mount below. L4/4/1/2{b:J3/3<f:M13>. Other manuals: Mosin Type VI 2636 (sim, but without star)

1580 *ca; [Pozzuoli]; Rome, Nicolas van Aelst; Tooley 459, NL 4F345.*

183. Bollard in four parts in circle A; 42 mm dia., with hooked top-piece pointing R, and narrow base (16.5 mm).

[**1569**]; [Geneva]; n.pl., Donatum Bertelli; n.i. Tooley, CSUF ac 1329.

***1569** ca; Constantinopoli; [Venice], Giovanni Francesco Camocio; n.i. Tooley, GHB 362.*

184. Bollard in four parts in circle B; 42 x 44 mm, with hooked curleque, curved body, and splayed base (18 mm wide).

*[**1569**]; [Jerusalem]; n.pl., Donatum Bertelli; n.i. Tooley, CSUF 1330.*

183

184

185

185. Wheel (41 x 42mm) with eight spokes A. L5/4. Other manuals: BEANS 11 (sim.)

1558; [Holland]; n.pl., n.pr.; TOOLEY 299, NL VCA 2.

1562; [Africa]; Venice, Paolo Forlani; TOOLEY 67, NL VCA 11(i).

1562; [Africa]; Venice, Paolo Forlani; TOOLEY 67, NL VCA 1.

1563; [Rome]; Venice, Ferrando Bertelli; TOOLEY 480, Doria Atlas 69.

1563; [Rome]; Venice, Ferrando Bertelli; TOOLEY 480, NL VCA 4.

1563; [Tuscany]; Venice, Ferrando Bertelli (Paolo Forlani); TOOLEY 574, NL VCA 30.

1564; [Germany]; Venice, Ferrando Bertelli (Paolo Forlani); TOOLEY 256, NL VCA 35.

1564; [Greece]; Venice, Ferrando Bertelli (Domenico Zenoi); TOOLEY 286, NL VCA 12.

1565; [Africa]; n.pl., Ferrando Bertelli; TOOLEY 73(a), NL VCA 38.

1565; [Algiers]; n.pl., Luca Bertelli; TOOLEY 98, NL VCA 27.

1565; [Austria]; n.pl., Ferrando Bertelli; TOOLEY 123, NL VCA 34.

1565; [Dalmatia]; Venice, Ferrando Bertelli; TOOLEY 188, NL VCA 18.

1566; [Austria]; Venice, Paolo Forlani; TOOLEY 124, NL 4F176.

1566; [Frisia]; Venice, Jacob von Deventer (Giovanni Francesco Camocio); TOOLEY 219 (but with "Antiovissimae" added), GHB 121.

***1566**; [Palestine]; Venice, Paolo Forlani; TOOLEY 436, NL VCA 19.*

1566; [Switzerland]; Venice, Vincenzo Lucchino; TOOLEY 538, NL 4F163r.

n.d./1564; [Hispañola/Cuba]; Venice, Paolo Forlani; TOOLEY 89/84, NL VCA 24/25.

n.d. [Regno di Napoli]; n.pl., n.pr.; TOOLEY 404, NL VCA 8.

186. Wheel (40 x 41mm) with eight spokes B. L5/4. Other manuals: BEANS 11 (sim.)

1560; [Atlantic Ocean]; Venice, Paolo Forlani; TOOLEY 77, NL VCA 41.

1562; [Africa]; Venice, Paolo Forlani; TOOLEY 67, NL VCA 11(ii).

1562; [Cyprus/Crete]; Rome, Ferrando Bertelli (Mario Cartaro)/Venice, Ferrando Bertelli (Mario Cartaro); TOOLEY 182/175, NL VCA 22/23.

1562; [Germany]; n.pl., Ferrando Bertelli; TOOLEY 249, NL VCA 37.

1562; [Savoy]; Venice, Ferrando Bertelli; TOOLEY 513, NL VCA 28.

1564; *[Friuli]; Venice, Paolo Forlani; TOOLEY 229, NL VCA 32.*

1564; [Greece]; Venice, Ferrando Bertelli (Domenico Zenoi); TOOLEY 286, NL VCA 12.

1565; [Africa, Arabia, India]; n.pl., Ferrando Bertelli; TOOLEY 73b, NL VCA 20.

1565; [Africa, Arabia, India]; n.pl., Ferrando Bertelli; TOOLEY 73c,
NL VCA 21.

1565; [Belgium]; Rome, Ferrando Bertelli; TOOLEY 135, NL VCA 36.

1565; [Lombardy]; Venice, Ferrando Bertelli; TOOLEY 353, NL 4F307.

1566; [Bavaria]; Venice, Paolo Forlani; TOOLEY 128, NL VCA 33.

1566; [Egypt]; Venice, Paolo Forlani; TOOLEY 193, NL VCA 40.

n.d. [South America]; Venice, Paolo Forlani; TOOLEY 93, NL VCA 3.

n.d. [Tokay]; n.pl., Domenico Zenoi; TOOLEY 550, NL VCA 5.

n.d. [Transylvania]; n.pl., Domenico Zenoi; TOOLEY 552, NL VCA 13.

n.d. [Malta]; n.pl., Domenico Zenoi (Luca Bertelli); n.i. TOOLEY, NL VCA 14.

n.d. [Sicily]; n.pl., Ferrando Bertelli; TOOLEY 521, NL VCA 29.

186

187

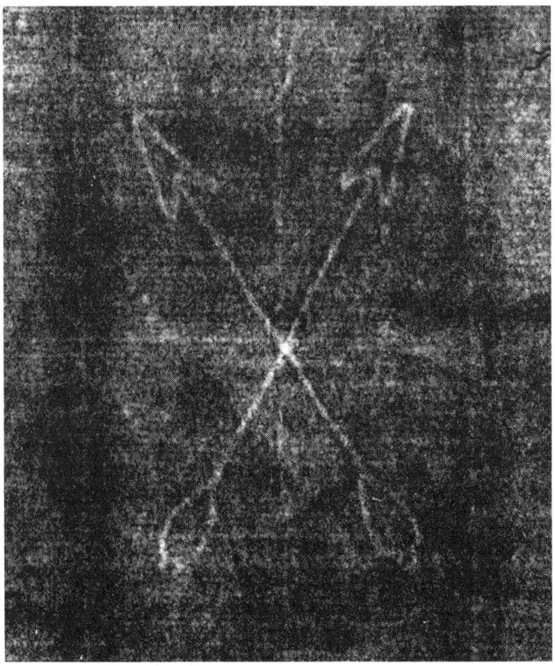

188

187. Arrows crossed A; (measurements: distance in mm between points, length of first shaft [top half], length of second shaft [top half], angle between shafts – 34, 36, 36, 60°); 2 chainlines 49 mm. (2*(x:M1))

1552; [Greece]; Rome, N. Sophianus; TOOLEY 278, GHB 144r.

1552; *[Greece]; Rome, N. Sophianus; TOOLEY 278, GHB 144v.*

188. Arrows crossed B; (measurements: distance in mm between points, length of first shaft [top half], length of second shaft [top half], angle between shafts – 38, 36, 38, 60°); 2 chainlines 54 mm. (2*(x:M1)). Other manuals: HEAWOOD 1929 10 (sim.)

1558; *[Holland]; n.pl., n.pr.; TOOLEY 299, NL 4F141.*

189. Arrows crossed C; (measurements: distance in mm between points, length of first shaft [top half], length of second shaft [top half], angle between shafts – 44, 44, 44, 60°); one feather 7 mm wide. (2*(x:M1))

***1567**; [Switzerland]; Venice, Paolo Forlani;* Tooley *539, NL 4F164.*

190. Arrows crossed D; (measurements: distance in mm between points, length of first shaft [top half], length of second shaft [top half], angle between shafts – 40, 37, 37, 65°); one feather tear shaped, the other larger. (2*(x:M1))

1552; [Greece]; Rome, N. Sophianus; Tooley 278, GHB 144r.

***1552**; [Greece]; Rome, N. Sophianus;* Tooley *278, Boswell GHB 144v.*

189

190

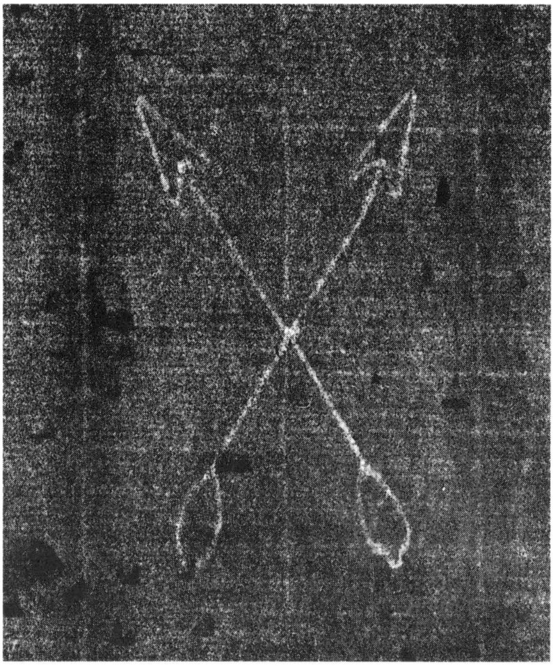

191

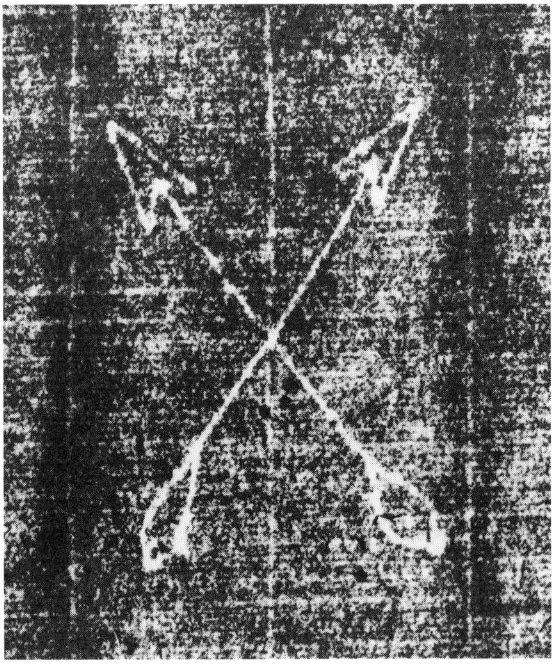

192

191. Arrows crossed E; (measurements: distance in mm between points, length of first shaft [top half], length of second shaft [top half], angle between shafts – 39, 36, 37, 65°); one feather almond shaped, offset to center, the other has clear notch at end. (2*(x:M1))

1555; [Flanders]; Venice, Michele Tramezzino (Jacob Bos), TOOLEY 198; NL 4F150.

***1556**; [Lombardy]; Rome, Vincenzo Lucchino; TOOLEY 346, NL 2F156.*

1558; [British Isles]; Rome, Sebastiano di Re; TOOLEY 271, Doria Atlas 16.

1558; [Frisia]; Venice, Michele Tramezzino (Jacob Bos); TOOLEY 218, NL 4F145.

192. Arrows crossed F; (measurements: distance in mm between points, length of first shaft [top half], length of second shaft [top half], angle between shafts – 43, 36, 37, 70°); rounded feathers, one wider than other. (2*(x:M1)). Other manuals: BRIQUET 6282 (sim.); BEANS 4; BORRONI SALVADORI 38

1558; [Belgium]; Venice, Michele Tramezzino; TOOLEY 143, NL 4F154.

1558; [France]; Rome, Michele Tramezzino (Pirro Ligorio, Oronce Finé, Sebastiano di Re); TOOLEY 208, GHB EFT 65180.

***1558**; [Holland]; n. pl., Michele Tramezzino; TOOLEY 299, GHB Bg7*

1559; [Hungary]; Rome, Michele Tramezzino (Sebastianus Clodiensis); TOOLEY 308, NL 4F182.

n.d. [Brabant]; n.pl., n.pr.; n.i. TOOLEY, CSUF 14418.

193. Arrows crossed under star A; (measurements: distance in mm between points, length of first shaft [top half], length of second shaft [top half], distance from center to top of star, angle between shafts – 33, 38, 38, 60, 50°). (2*(x:M1)){t:J5/4}

1563; [Belgium]; n.pl., n.pr.; TOOLEY 134, NL 2F84.

1568; *[Augsburg]; Venice, Luca Guarinoni; TOOLEY 115, NL 4F173.*

194. Arrows crossed under star B; (measurements: distance in mm between points, length of first shaft [top half], length of second shaft [top half], distance from center to top of star, angle between shafts – 45, 35, 39, 55, 75°). (2*(x:M1)){t:J5/4} Other manuals: BEANS 3 (sim.)

1558; [Naples]; Rome, Michele Tramezzino; TOOLEY 403; Doria Atlas 73.

1561; [Portugal]; Rome, Michele Tramezzino (Sebastianus Clodiensis); TOOLEY 456, NL 4F195.

1561; *[Portugal]; Rome, Michele Tramezzino (Sebastianus Clodiensis); TOOLEY 456, NL 4F196.*

1563; [Friuli]; Rome, Michele Tramezzino; TOOLEY 230, Doria Atlas 64.

193

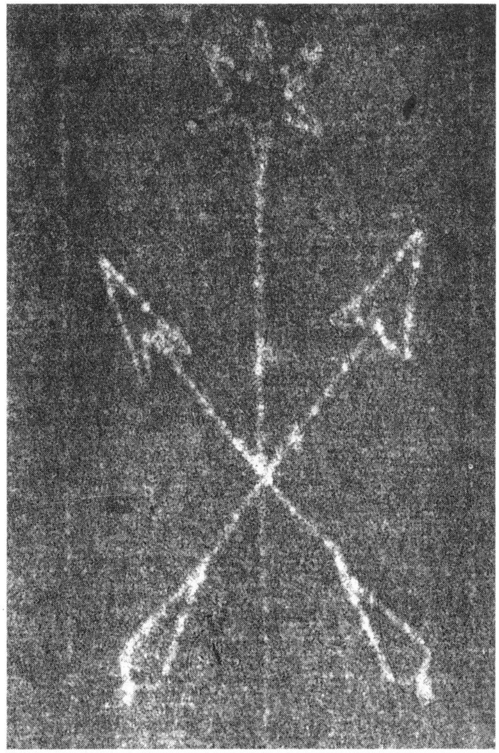

194

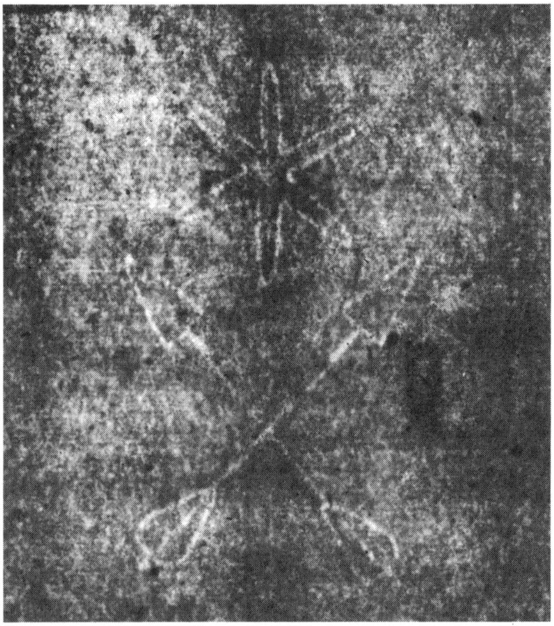

195

195. Arrows crossed under star C; (measurements: distance in mm between points, length of first shaft [top half], length of second shaft [top half], distance from center to top of star, angle between shafts – 40, 30, 30, 48, 85°. (2*(x:M1)){t:J5/4}

1546; *[World]; Venice, Giacomo Gastaldi;* T*OOLEY 3, NL 4F4.*

196

196. Arrows crossed under star D; (measurements: distance in mm between points, length of first shaft [top half], length of second shaft [top half], distance from center to top of star, angle between shafts – 41, 40, 40, 60, 60°). (2*(x:M1)){t:J5/4} Other manuals: BEANS 29 (possible)

1566; [Greece]; Venice, Paolo Forlani (Giovanni Francesco Camocio); TOOLEY 287, GHB 179r.

1566; *[Greece]; Venice, Paolo Forlani (Giovanni Francesco Camocio); TOOLEY 287, GHB 179v.*

197. Arrows crossed under star E; (measurements: distance in mm between points, length of first shaft [top half], length of second shaft [top half], distance from center to top of star, angle between shafts – 35, 31, 33, 60, 70°). (2*(x:M1)){t:J5/4}

1552; *[Rome]; Rome, Michele Tramezzino (Pirro Ligorio); Tooley 484, NL 4F246.*

1580; [Palermo]; Rome, Claudio Duchetti (Natale Bonifacio); Tooley 431, NL 4F238.

198. Arrows crossed under star F; (measurements: distance in mm between points, length of first shaft [top half], length of second shaft [top half], distance from center to top of star, angle between shafts – 43, 36, 37, 80°); shaft bent just above left feather. (2*(x:M1)){t:J5/4}

1558; *[Naples, Kingdom]; Rome, Sebastianus Clodiensis (Pirro Ligorio); Tooley 403, NL 4F340.*

1564; [Friuli]; Rome, Michele Tramezzino (Pirro Ligorio); Tooley 230, NL 4F302.

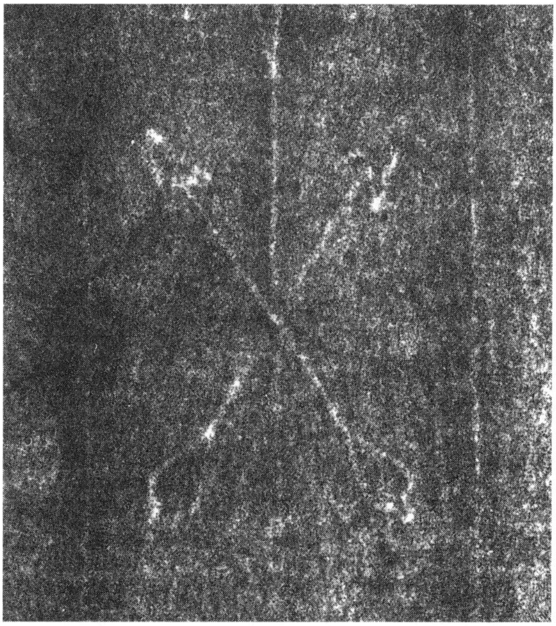

197

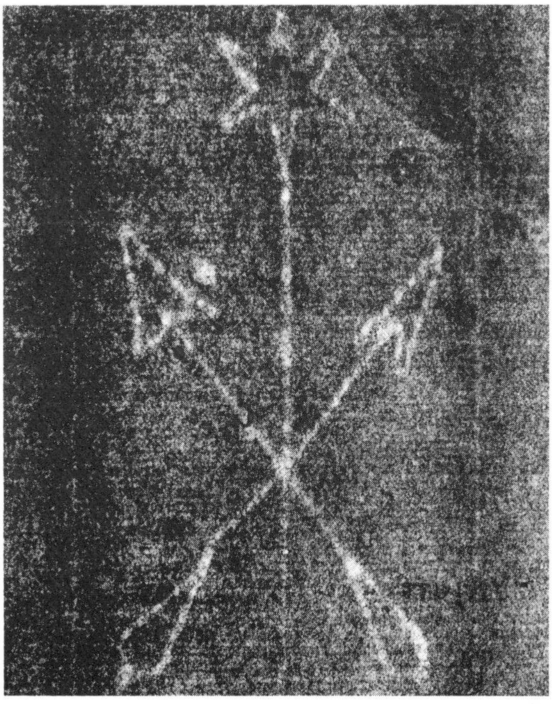

198

199. Arrows crossed under star G; (measurements: distance in mm between points, length of first shaft [top half], length of second shaft [top half], distance from center to top of star, angle between shafts – 34, 35, 35, 57, 60°); star is four-pointed. (2*(x:M1)){t:J5/2}. Other manuals: ZONGHI 1248

1549; [*Rome (Palazzo Farnese)*]; *Rome, Antonio Lafreri; n.i.* TOOLEY, *NL 4F261.*

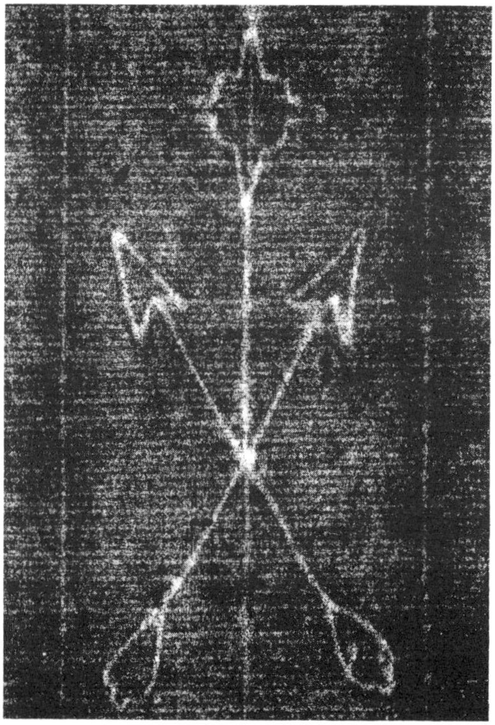

199

200. Arrows crossed under star H; (measurements: distance in mm between points, length of first shaft [top half], length of second shaft [top half], distance from center to top of star, angle between shafts – 23, 37, 36, 53, 56°). (2*(x:M1)){t:J5/2} Other manuals: BEANS 1 (sim.)

1575; [*Zara*]; *n.pl., n.pr.;* TOOLEY *601 (later copy), NL 4F104.*

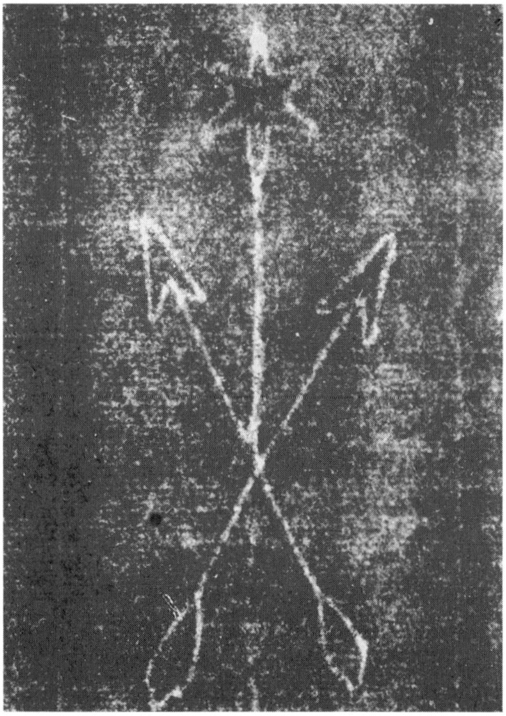

200

201. Arrows crossed under star with downward points; measurements: distance between feathers 42 mm, length of first shaft [top half, feather] 34 mm, length of second shaft [top half, feather] 34 mm, distance from center to top of star 65 mm, angle between shafts 70°. (2*(x:M1)){t:J5/4}. Other manuals: MASSARI (1983) 226 (sim.).

***1570** ca; [Lombardy]; Venice, Ferrando Bertelli; n.i. TOOLEY, NL 4F308.*

202. Double-bow and arrow in single wire (57 x 42mm). M3[M1]

***n.d.** [Tuscany]; Venice, Ferrando Bertelli; TOOLEY 577, GHB 842.*

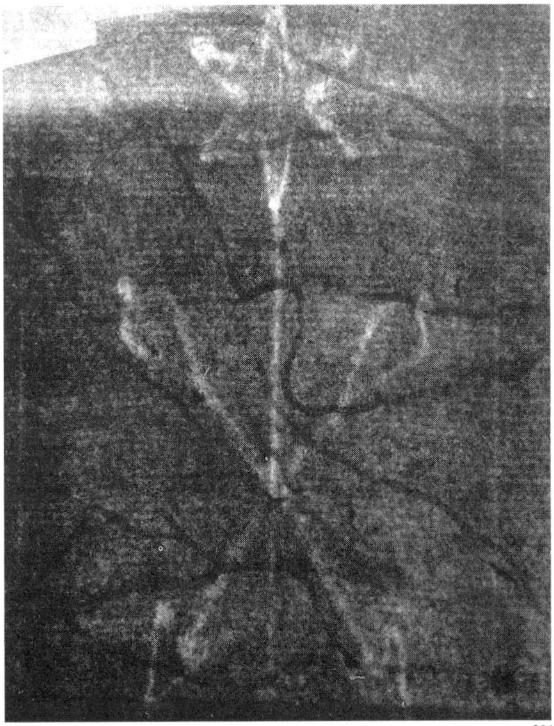

201

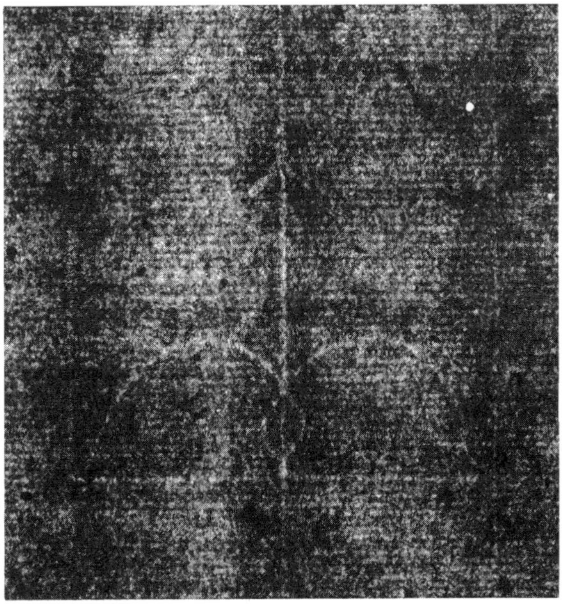

202

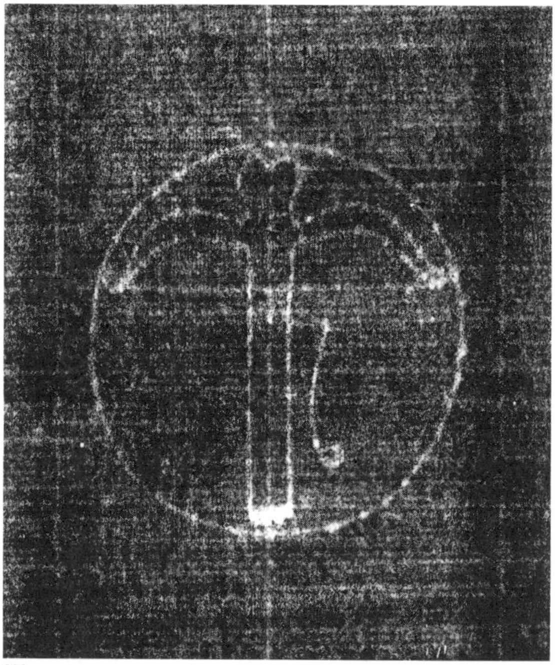

203

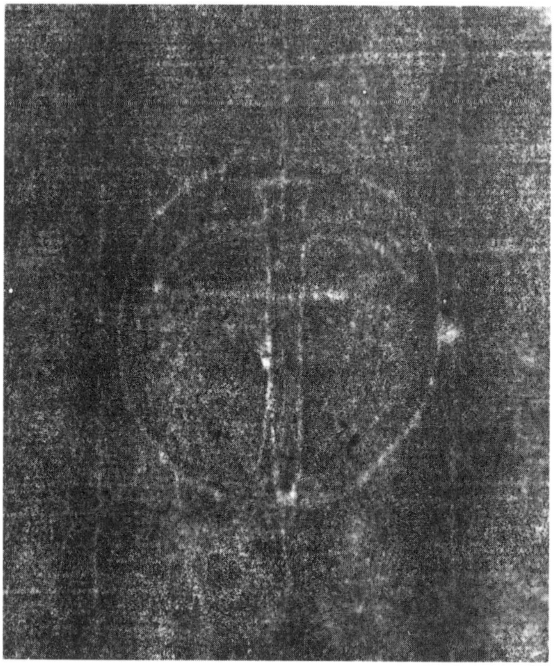

204

203. Crossbow in circle A; 51 x 52 mm. M6<f:U1>. Other manuals: BRIQUET 749 (sim.)

1566; [Venice]; Venice, Paolo Forlani; TOOLEY 582, NL VCA 17.

__1566__; [Venice]; Venice, Bolognino Zaltieri (Paolo Forlani); TOOLEY 586, NL VCA 39.

1566; [Venice]; Venice, Paolo Forlani (Bolognino Zaltieri); n.i. TOOLEY, NL 4F290.

204. Crossbow in circle B; flathead crossbow; circle 44 x 46 mm. M6<f:U1>

__1556__; [Piedmont]; Venice, Gabriel Giolito (Fabio Licinio, Giacomo Gastaldi); TOOLEY 448, NL VCA 9.

205. Crossbow in circle C; flathead crossbow; circle 43 x 48 mm; cocking handle with no circle at end. M6 <f:U1>

1559; [*Hungary*]; *Rome, Michele Tramezzino (Wolfgang Lazius, Pirro Ligorio); TOOLEY 307, KN 096-050-21.*

206. Crossbow in circle D; circle 53 x 44 mm; sim to #211. M6<f:U1>

n.d. [*Corsica*]; *n.pl., Fabio Licinio; TOOLEY 169, NL 2F80.*

205

206

207

208

207. Crossbow in circle E; circle 45mm dia; sim to #210. M6<f:U1>

1557; [Rome]; Rome, Antonio Lafreri (Nicholas Beatrizet); TOOLEY 472, KN 051-21-#2.
n.d. *[Szighet]; Rome, n.pr.; TOOLEY 546, GHB 228.*

208. Crossbow in circle F; circle 45 mm dia; large cocking handle; sim to #213. M6<f:U1>

1556; [Great Britain]; n.pl., n.pr. (George Lily orig.); TOOLEY 269, NC 1:48:3.Other manuals: ROBERTS – Crossbow B
1558; *[Calais]; n.pl., n.pr.; TOOLEY 152, NL 4F96.*

209. Crossbow in circle G; circle 45 x 46 mm; large, but bent, cocking handle; sim to #212. M6<f:U1>

1560; *Il vero disegno della Fortezza di Sziget; Rome, n.pr.; Tooley 546, NL 2F95.*

210. Crossbow in circle H; circle 46 mm dia; long, flat cocking handle. M6<f:U1>

1560; [*Atlantic Ocean*]; *Venice, Giovanni Francesco Camocio (Venice, Paolo Forlani); Tooley 77,* NC *1:48:5 right.*

1560; [Lombardy]; Venice, Giovanni Francesco Camocio; Tooley 349, NC 1:48:18.

209

210

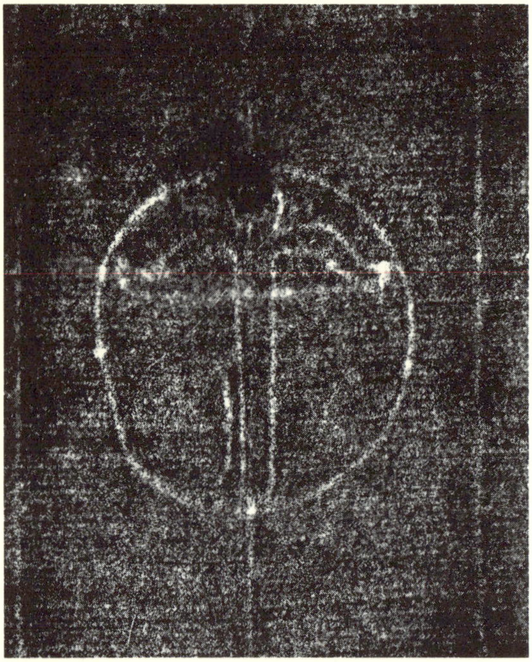

211

211. Crossbow in circle I; circle 43 x 46, crooked; long cocking handle. M6<f: U1>. Other manuals: BEANS 25 (sim.)

n.d. *[Rome]; n.pl., n.pr.; [*TOOLEY *478],* NC *1:48:21.*

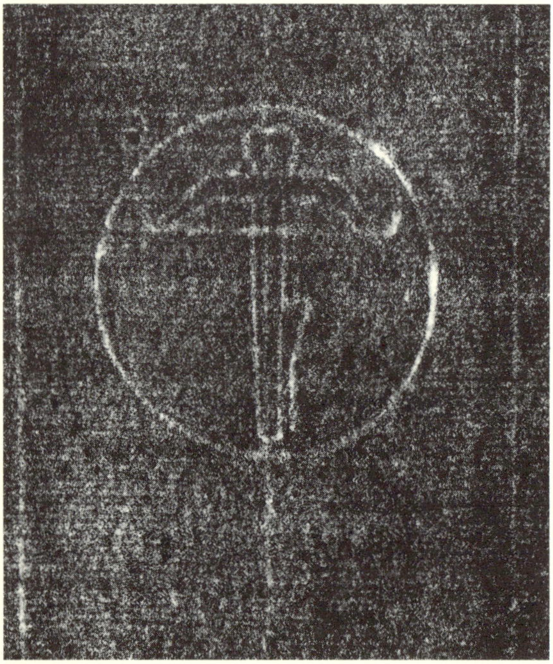

212

212. Crossbow in circle J; circle 46 mm dia. M6<f:U1>

n.d. *[Ireland]; n.pl., n.pr.;* TOOLEY *315,* NC *1:48:29.*

213. Crossbow in circle K; circle 48 x 50 mm. M6<f:U1>

***1540** ca; [Piedmont]; n.pl., Matteo Pagano; n.i. Tooley, NL 4F312.*

214. Crossbow in circle under six-pointed star A; circle 41 x 46 mm; star 24 mm tall; total height 70 mm. M6<f: U1{t:J5/4}>

***1572**; [Battle of Lepanto]; Brescia, Stefano Ghibellini; n.i. Tooley, NL 4F109.*

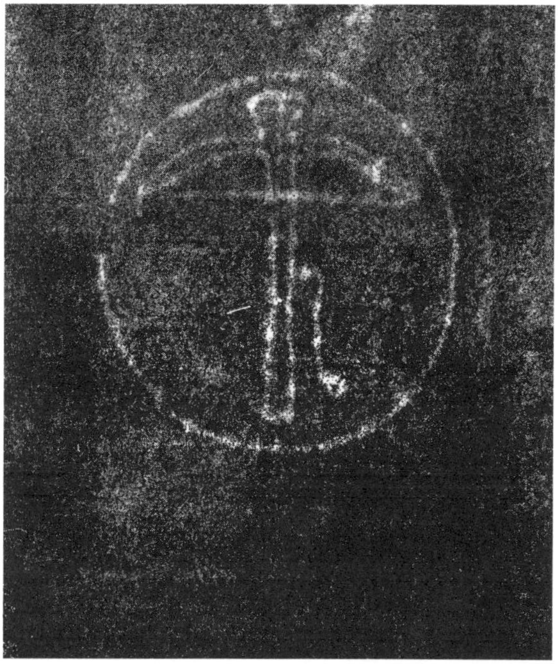

213

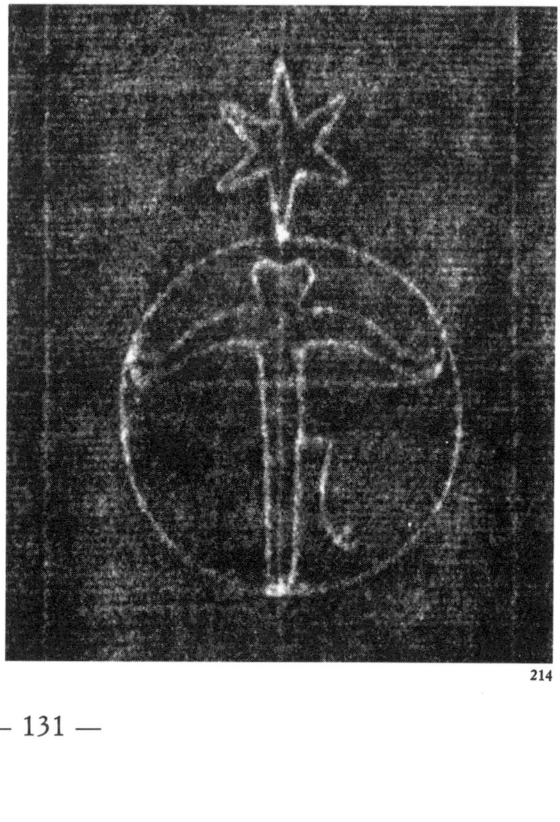

214

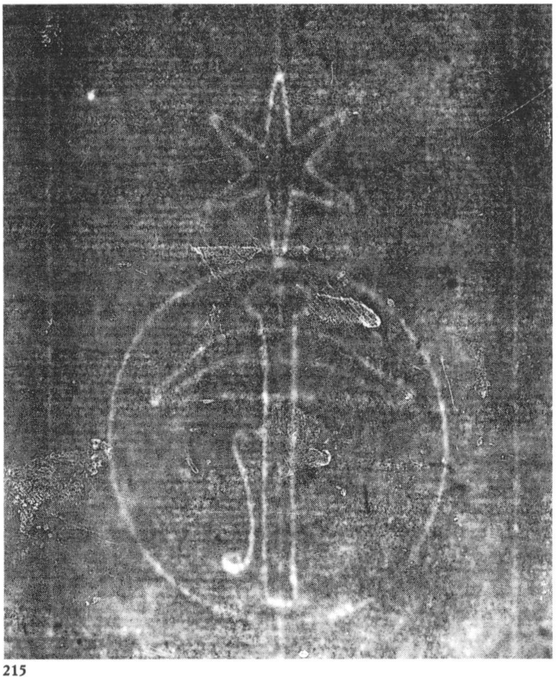

215

216

215. Crossbow in circle under six-pointed star B; circle 46 x 47 mm; star 25 mm tall; total height 72 mm. M6<f: U1{t:J5/4}>

1557; *[Palestine]; Rome, Joannem Franciscium della Gatta (Basle, Wissenburg orig.); Tooley 434,* NC *2:130:60.*

216. Crossbow in circle under fleur-de-lys; flathead crossbow; circle 48 x 50 mm; fleur-de-lys 21 mm tall; total height 71 mm. M6<f:U1{t:G6/2}>. Other manuals: Roberts – Crossbow D (sim.)

1542; *[Europe (France, Spain, Flanders]; n.pl., Aeneas Vico; n.i. Tooley, NL 4F131.*

1548; [Palestine]; n.pl., Antonio Salamanca; Tooley 433, NL 2F229.

1551; [Malta]; Rome, Antonio Lafreri; Tooley 373, NC 1:48:25.

1554; [France, Spain, Flanders]; n.pl., n.pr.; Tooley 207, NC 1:48:1.

217. Crossbow in circle with a loaded arrow A; flathead crossbow; circle 45mm dia; arrow 27mm tall; total height 72 mm; cocking handle touches stock. M6<f:U1[M1]>. Other manuals: BRIQUET 759 (sim.)

1551; *[Malta]; Rome, Antonio Lafreri;* TOOLEY *373, NL 2F181.*

218. Crossbow in circle with a loaded arrow B; circle 49 x 50 mm; arrow 35 mm tall; total height 85 mm. M6<f:U1[M1]>

1558; *[Belgium]; Rome, Michele Tramezzino (Sebastianus di Re);* TOOLEY *132,NL 4F147.*

217

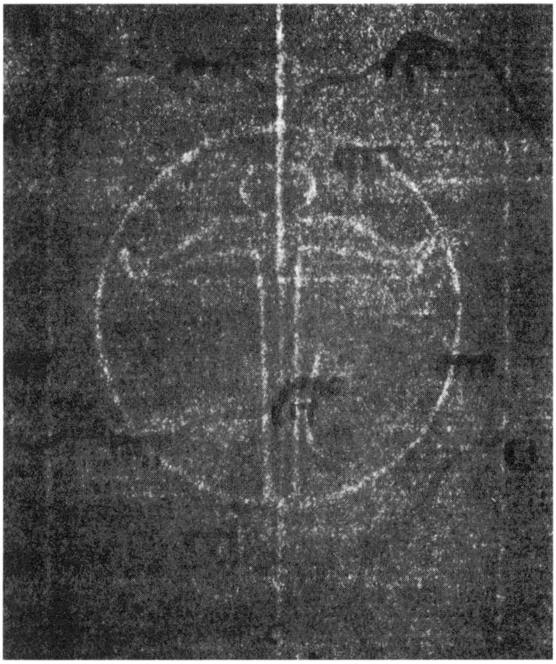

218

219. Crossbow in circle with a loaded arrow C; circle 49 x 51 mm; total height 87mm. M6<f:U1[M1]>

n.d. *Il disegno del paese dove se trovano i due eserciti de sua Maiesta cesarea; n.pl, Tomasso B.; n.i.* TOOLEY, NC *1:48:41.*

220. Crossbow in circle with a loaded arrow D; circle 49 x 51 mm; total height 88mm. M6<f:U1[M1]>

1542; [Budapest]; n.pl., E[nea] V[ico]; TOOLEY 150, NC 1:48:43.

n.d. *[Ingolstadt]; Venice, Tom[asso] B.;* TOOLEY *314,* NC *1:48:40.*

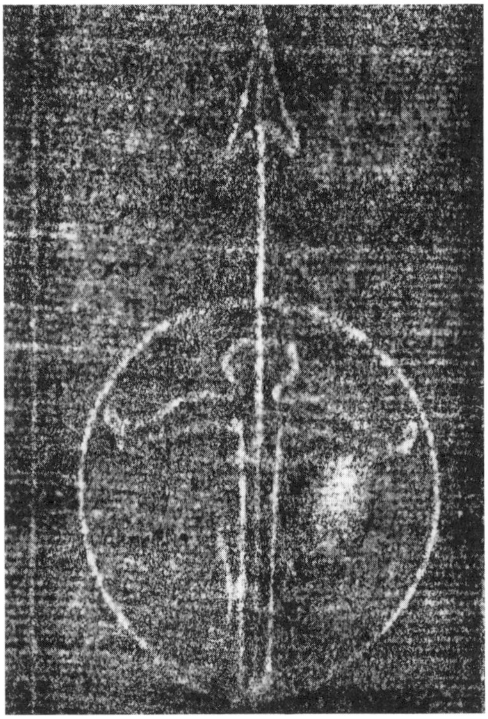

219

220

221. Shield with three chevrons above a wolf in circle under a six-pointed star A; circle 43 x 45 mm; total height 70 mm. M13{b:C7<f:U1{t:J5/4}>}

***1560** ca; [Genoa]; n.pl., Giovanni Francesco Camocio; Tooley 240, NL 4F222.*

222. Shield with three chevrons above a wolf in circle under a six-pointed star B; circle 47 x 46 mm; total height 69 mm. M13{b:C7<f:U1{t:J5/4}>}

1562; [Germany and neighboring provinces]; Rome, Antonio Lafreri (Ferrando Bertelli orig.); Tooley 249, NC 2:130:25 right.

1566; [Asia Minor]; Rome, Antonio Lafreri (Venice, Giovanni Francesco Camocio orig.); Tooley 65, NC 2:130:56 left.

1566; [Asia Minor]; Rome, Antonio Lafreri (Venice, Giovanni Francesco Camocio orig.); Tooley 65, NC 2:130:56 right.

1566; [New France]; Rome, Antonio Lafreri (Venice, Bolognino Zaltieri orig.); Tooley 67, NC 2:130:67.

1570; [Cyprus]; Venice, Natale Bonifacio; Tooley 187, NL 2F222.

***1570**; [Bavaria]; Venice, Paolo Forlani; Tooley 130, NL 2F91.*

n.d. [Bohemia]; Rome, Antonio Lafreri; Tooley 138, NC 2:130:26 left.

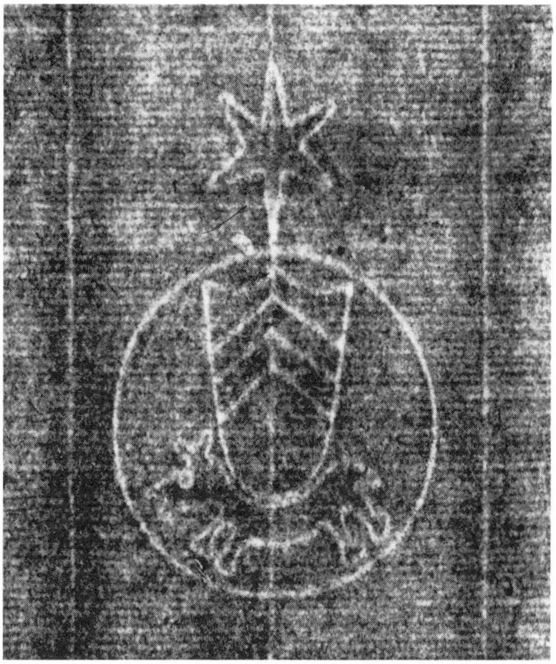

221

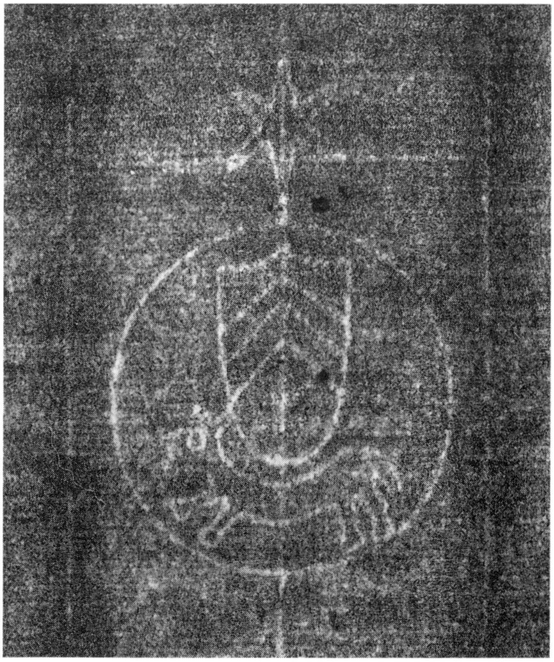

222

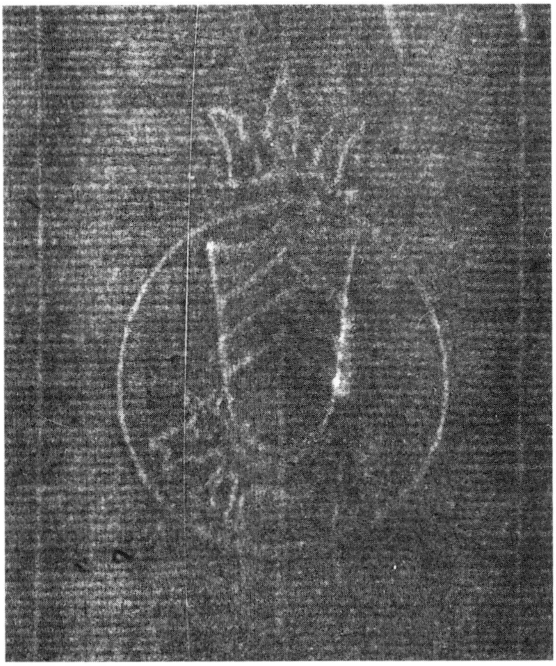

223

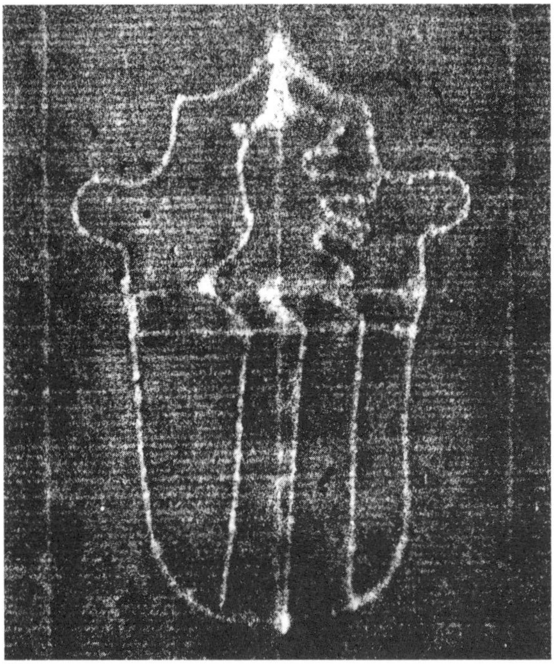

224

223. Shield with three chevrons above a wolf in circle under a six-pointed star C; circle 45 x 46 mm; crown with three points; total height 64 mm. M13 {b:C7<f:U1{t:R3}>}

*[**1570**]; [Grand Duke of Tuscany's arrival in Rome, 18 February 1570]; Rome, Antonio Lafreri; n.i. TOOLEY, NL 4F263.*

224. Shield with unicorn A; shield 54 x 80 mm with knobs at sides, containing top half of a unicorn with two bars beneath. M13[F4]. Other manuals: BEANS 61 (sim.)

1565; [Malta]; Rome, Antonio Lafreri; TOOLEY 374, Boswell 246r.

***1566**; [Valletta]; Rome, Antonio Lafreri; TOOLEY 361, Boswell 214r.*

1585; [Ancona]; Rome, Claudio Duchetti (Ambrogio Brambilla); TOOLEY 109, Boswell 233r.

225. Shield with unicorn B; shield 49 x 76 mm with knobs at sides, containing top half of a unicorn with two bars beneath. M13[F4] Other manuals: BEANS 61 (sim.)

1577; [Italy]; Rome, Mario Cartaro; TOOLEY 337, NL 4F206.

226. Shield with unicorn C; shield 48 x 80mm with the top half of a unicorn and two bars beneath. M13[F4]

1565; [Malta]; Rome, Antonio Lafreri; TOOLEY 284, Boswell 247r.

1566; [Zighet, Hungary]; Rome, Antonio Lafreri; TOOLEY 544, Boswell 248r.

1571; [Europe]; Venice, Paolo Forlani (Claudio Duchetti); TOOLEY 39, NL 4F91v.

1571; [Europe]; Venice, Paolo Forlani (Claudio Duchetti); TOOLEY 39, NL 4F91r.

1581; [Genoa]; Rome, Claudio Duchetti (Ambrogio Brambilla); TOOLEY 243, Boswell 231r.

1585; [Naples]; Rome, Claudio Duchetti; TOOLEY 412, Boswell 230r.

1585 ca; [Venice]; Rome, Claudio Duchetti (Jacomo Franco); TOOLEY 585, Boswell 216r.

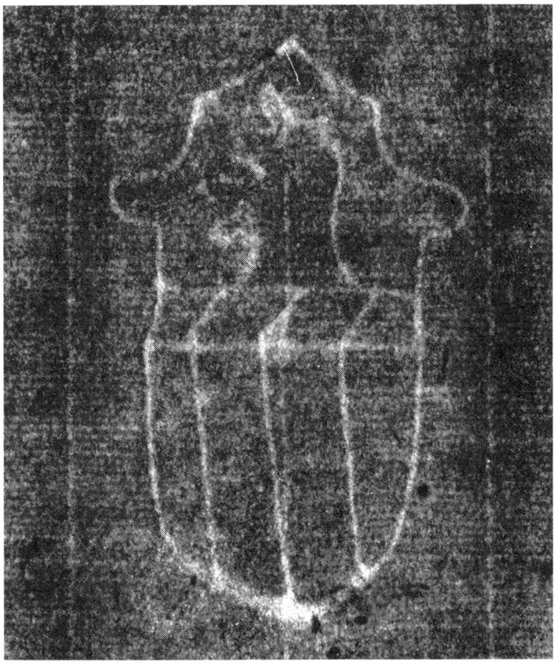

225

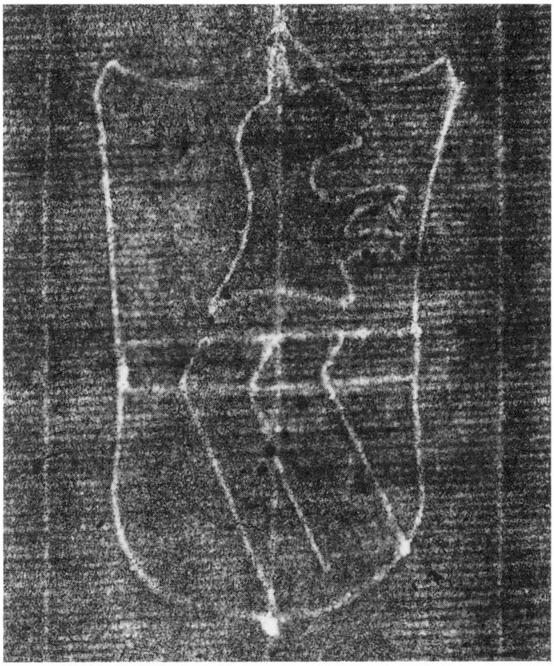

226

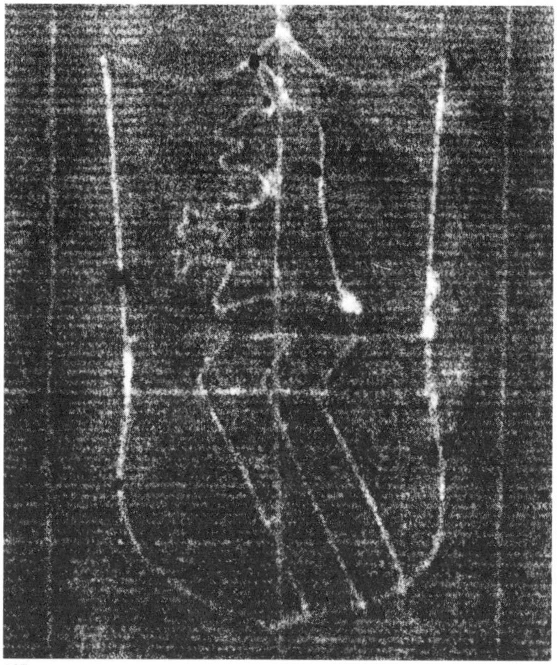

227

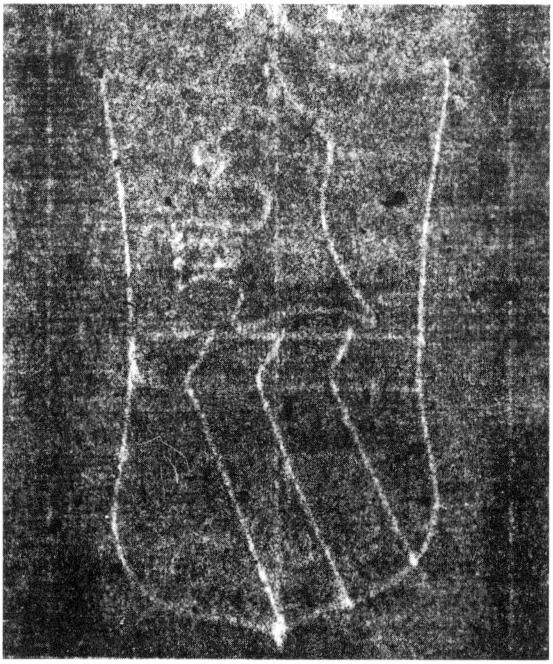

228

227. Shield with unicorn D; shield 46 x 79 mm with the top half of a unicorn and two bars beneath. M13[F4]

1550 ca; [Lake Geneva]; n.pl., Domenico Alphane; TOOLEY 238, Boswell 191r.

1582; *[Ancient Rome]; Rome, Ambrogio Brambilla (Claudio Duchetti); TOOLEY 494, Boswell 228r.*

1582; [Bologna]; Rome, Claudio Duchetti; TOOLEY 139, Boswell 238r.

1592; [Bichat]; Rome, Giovanni Battista de Cavalleri; TOOLEY 137, Boswell 249r.

228. Shield with unicorn E; shield 47 x 84 mm with the top half of a unicorn and two bars beneath. M13[F4]

1575; *[Rome, Villa di Tivoli]; Rome, Mario Cartaro; n.i. TOOLEY, NL 4F269.*

229. Shield with unicorn F; shield 48 x 86 mm with the top half of a unicorn and two bars beneath. M13[F4] Other manuals: BEANS 60

1585; *[Sicily]; n.pl, Mario Cartaro; n.i.* TOOLEY, *NL 2F176*
1586; [Ischia]; Naples, Mario Cartario; TOOLEY 320, NL 4F211.

230. Anvil and hammer in circle; 51 x 49 mm. N1-N12/1<f:U1>. Other manuals: BEANS 72 or 76 (sim.); HEAWOOD 1924 48 (sim.)

1574; *[Rome]; Rome, Etienne Dupérac; n.i.* TOOLEY, *NL 4F251.*

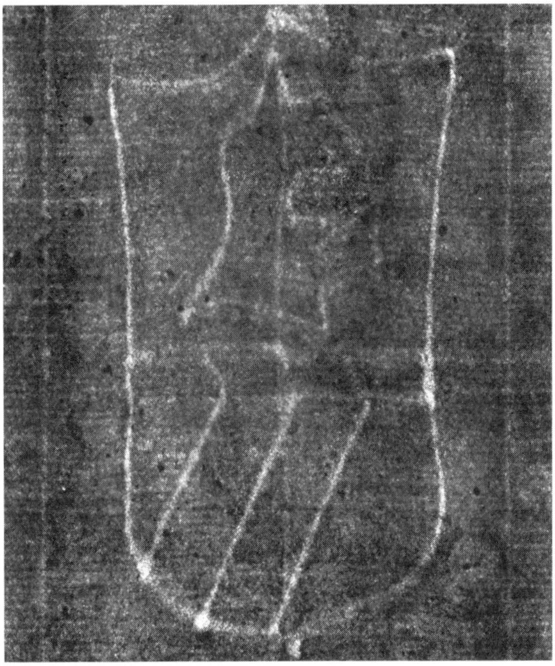

229

230

231

232

231. Anvil and hammer in circle under cross A; circle 48 x 51 mm; total height 73 mm. N1-N12/1<f:U1{t:S1}>. Other manuals: BEANS 69 or 75 (sim.); HEAWOOD 1924 49 (sim.); BRIQUET 5965 (sim.)

1564; [Rome, Vatican]; Rome, Vincenzo Lucchino; n.i. TOOLEY, NL 4F264.

1569; *[Rome, capitol]; Rome, Etienne Dupérac; n.i. TOOLEY, NL 4F266.*

232. Anvil and hammer in circle under cross B; circle 49 x 52 mm; total height 76mm. N1-N12/1<f:U1{t:S1}>. Other manuals: HEAWOOD 1924 49 (sim.)

1574; *[Rome, Tivoli]; Rome, Etienne Dupérac (Antonio Lafreri); n.i. TOOLEY, NL 4F268.*

233. Cardinal's hat A; simple top; brim 50 mm wide; ties 15 mm apart. N14. Other manuals: Zonghi 960 (sim.)

n.d. *[World, two hemispheres in gores]; n.pl., Antonio Floriano; Tooley 23, NL 4F46v.*

234. Cardinal's hat B; sides of top curved; brim 50 mm wide; ties 19mm apart. N14. Other manuals: Briquet 3395 (sim.)

1562; [Great Britain]; Rome, Antonio Lafreri (Venice, Ferrando Bertelli, orig.); Tooley 272, NC 2:130:[map 8 right margin].

n.d. *[World, two hemispheres in gores]; n.pl., Antonio Floriano; Tooley 23, NL 4F46r.*

n.d. [Estland]; Rome, Antonio Lafreri (orig?); Tooley 197, NC 2:130:5.

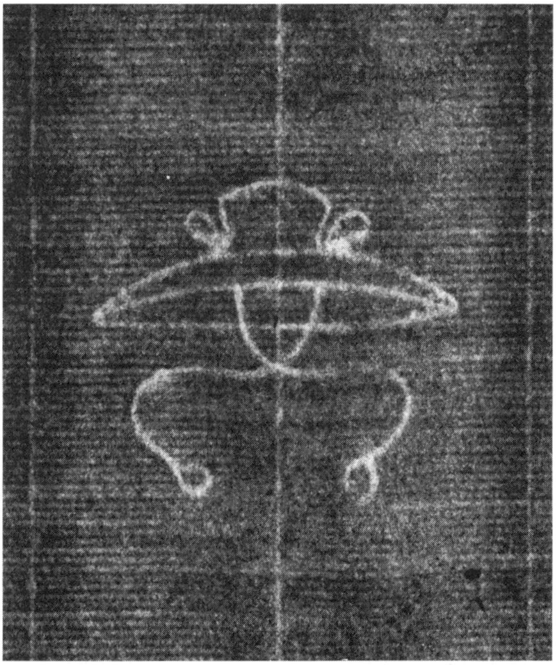

233

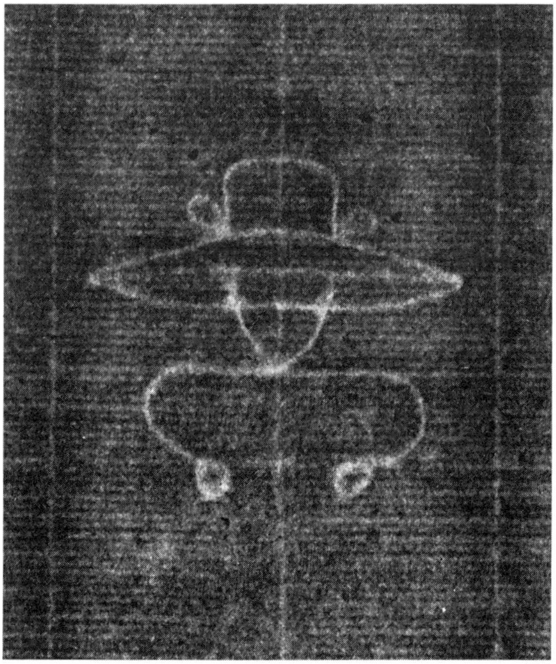

234

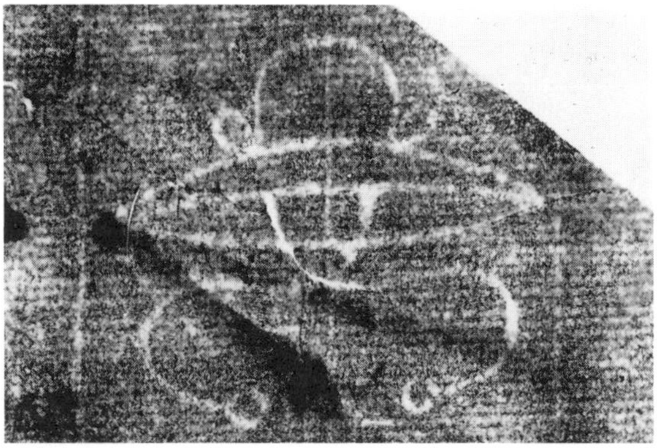

235

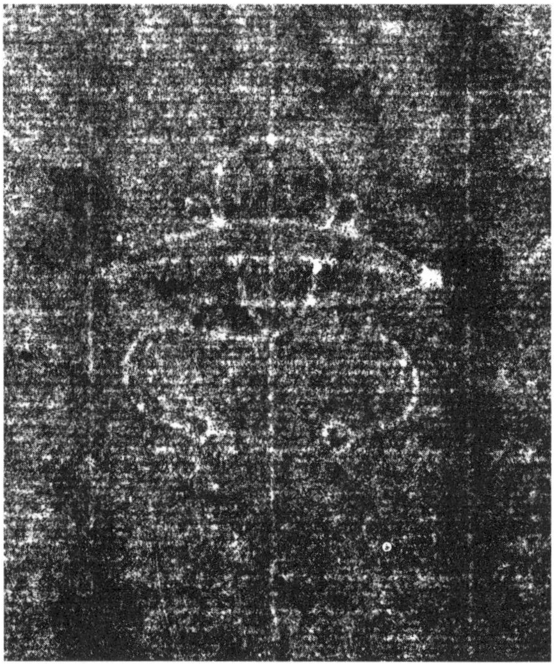

236

235. Cardinal's hat C; curved top; brim 49 mm wide; ties 16 mm apart. N14. Other manuals: Zonghi 960 (sim.)

1557; [*Palestine*]; *Rome, Franciscium Della Gatta; Tooley 434,* NC *1:48:31.*

236. Cardinal's hat D; curved top; brim 56 mm wide; ties 16 mm apart. N14. Other manuals: Briquet 3395 (sim.)

1558; [*Lombardy*]; *Rome, Vincenzo Lucchino; Tooley 347, NL 4F303.*

237. Cardinal's hat with trefoil; brim 41 mm wide; ties curved outwards; under trefoil (22 x 12 mm); total height 69mm. N14{t:G4}. Other manuals: BRIQUET 3415 (sim.); HEAWOOD 1924 69 (sim.)

1560; *[Greece]; Venice, Giacomo Gastaldi; TOOLEY 282, NL 2F185.*

238. Ladder in circle with six-pointed star A; three rungs in double wire (17 x 36); circle 43 x 44 mm under star; total height 71 mm. N18<f:U1{t:J5/4}>

1548; [Palestine]; Rome, Antonio Salamanca; TOOLEY 433, NC 2:130:59.

1548; *[Palestine]; Rome, Antonio Salamanca; TOOLEY 433, GHB 3069.*

1558 (after); [Gravelines]; n.pl., n.pr.; TOOLEY 604, KN 03-GES-21.

n.d. [Boulogne]; n.pl., n.pr.; [TOOLEY 140], NC 1:48:35.

n.d. [Antwerp]; n.pl., n.pr.; TOOLEY 110, NC 1:48:39 (bent prong on star).

237

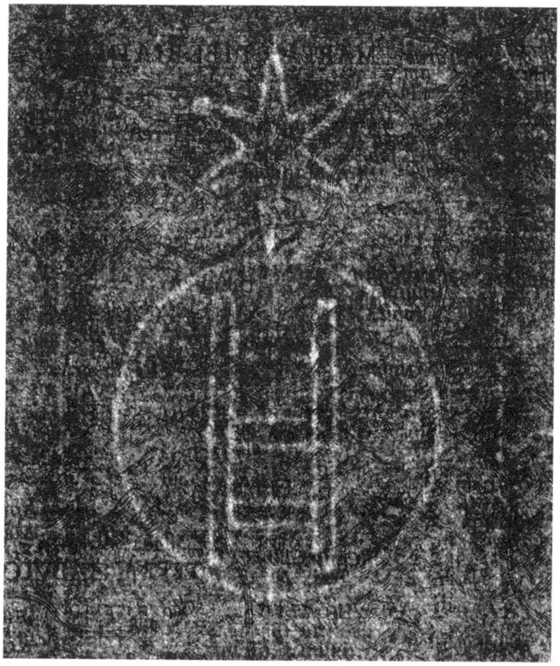

238

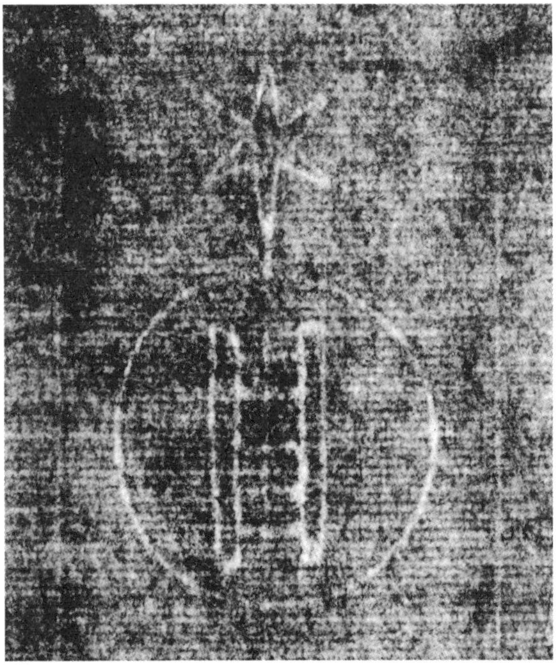

239

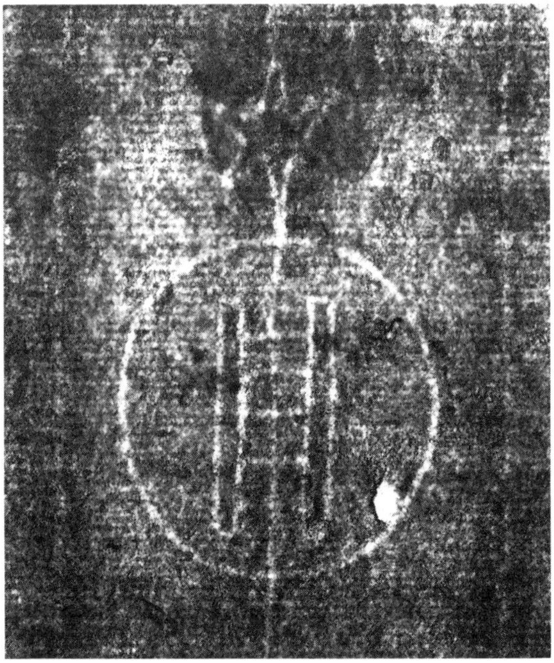

240

239. Ladder in circle with six-pointed star B; five rungs in single wire (15 x 32 mm); circle 43 mm dia, crooked; total height 71 mm; top of one pole of ladder is crushed. N18<f:U1{t:J5/4}>

1548; [Rome]; Rome, Antonio Salamanca; n.i. Tooley, NL 2F135.

1557; [Naples]; Venice, n.pr.; Tooley 401, Doria Atlas 72.

1558; [Rome, Festa del Tostaccio]; n.pl., n.pr.; n.i. Tooley, NL 2F137.

[**1558**]; [Gravelines]; n.pl., n.pr.; Tooley 604, KN 03-GES21 (ms).

1561; [Italy]; n.pl., n.pr.; Tooley 329, NL 4F201r.

1561; [Italy]; n.pl., n.pr.; Tooley 329, NL 4F201v.

1560; [World]; Venice, Giacomo Gastaldi (Paolo Forlani, Giovanni Francesco Camocio); Tooley 7, NL 4F6.

***1560**; [Lombardy]; Venice, Giovanni Francesco Camocio; Tooley 349, NL 2F157.*

n.d. [Jerusalem]; n.pl., Stefano Dupérac; Tooley 341, NL 2F232.

n.d. [Rome]; n.pl., n.pr.; Tooley 478, KN 08-ARN-15.

n.d. [Civitella]; n.pl., n.pr.; Tooley 154, NC 1:48:44.

n.d. [Mirandola] 405 x 308 mm; [Rome]., A[ntonio] S[alamanca]; n.i. Tooley, NC 1:48:48.

240. Ladder in circle with six-pointed star C; five rungs in single wire (16 x 32 m); circle 43 x 45 mm; total height 70mm. N18<f:U1{t:J5/4}>

***1557**; [Naples]; Venice, n.pr.; Tooley 401, NL 4F338.*

1560; [Iberian Peninsula]; Venice, n.pr.; n.i. Tooley, CSUF ac 1115 (missing right edge).

241. Ladder in circle with six-pointed star D; five rungs in single wire (16 x 31 mm); circle 43 x 44 mm; total height 67 mm. N18<f:U1{t:J5/4}>. Other manuals: Beans 5

1540 ca; [Mirandola]; n.pl., Antonio Salamanca; Tooley 394, NL 4F229.

1548; [Palestine]; n.pl., Antonio Salamanca; Tooley 433, NL 2F228.

1557; [France]; n.pl., n.pr.; n.i. Tooley (but sim. to Tooley 214, 215), NC 1:48:6.

242. Ladder in circle with six-pointed star E; five rungs in single wire (16 x 31 mm); circle 43 mm dia; bottom of one ladder pole is crushed; total height 71 mm. N18<f:U1{t:J5/4}>

1540 ca; [Tuscany]; n.pl., Antonio Salamanca; Tooley 567, NL 4F332.

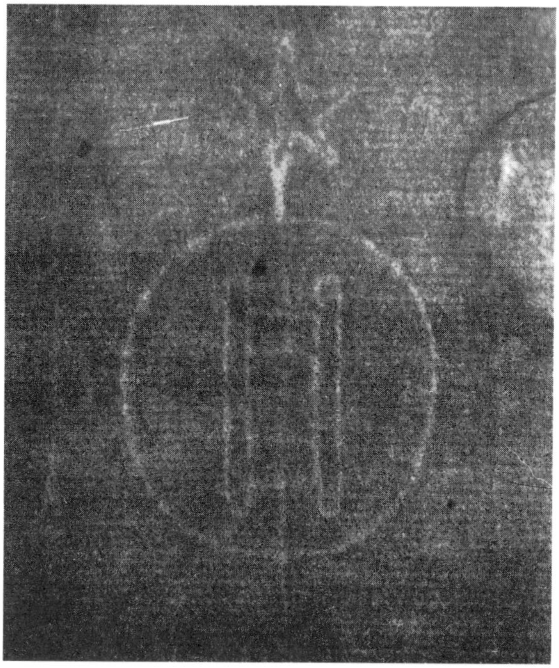

241

242

243. Ladder in shield under six-pointed star A; with four rungs (11 x 30 mm); shield 33 x 68 mm; total height 88 mm. N18<f:M13{t:J5/4}>

1557; *[Vicovaro]; n.pl., Sebastianus di Re; Tooley 594, NL 2F28.*

244. Ladder in shield under six-pointed star B; with three rungs (12 x 37); shield 30 x 75 mm; total height 92 mm. N18<f:M13{t:J5/4}>

1557; *[Vicovaro]; n.pl., Sebastianus di Re; Tooley 594, NL 2F28.*

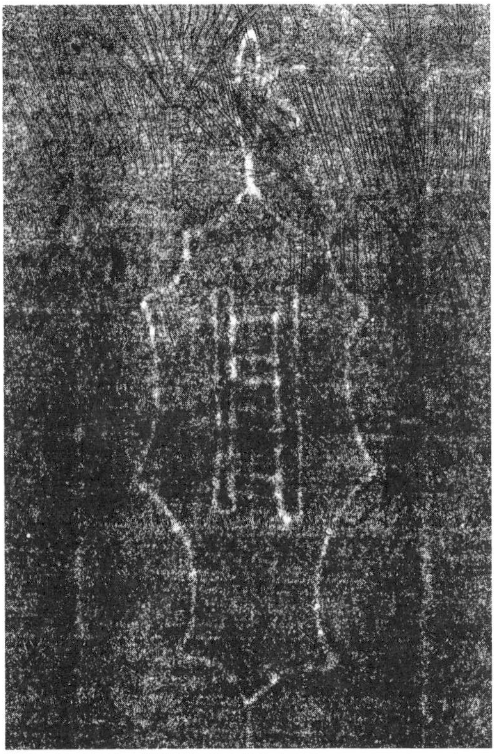

243

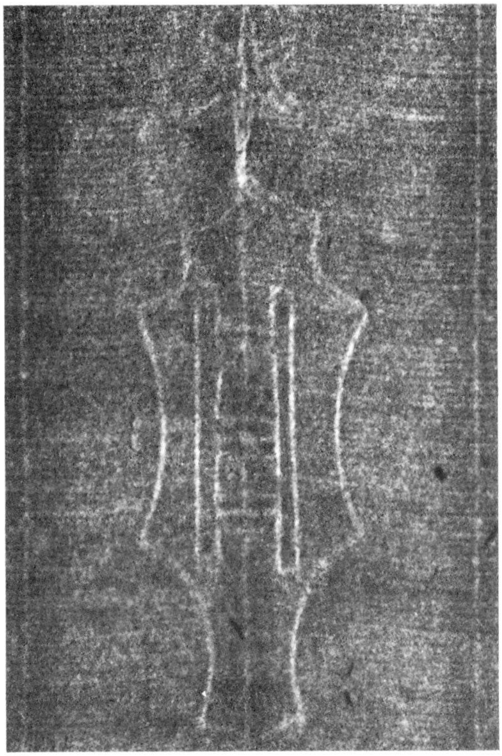

244

245. Ladder in shield under six-pointed star C; with three rungs (15 x 31 mm); shield 32 x 77 mm, pointed base; total height 100mm. N18<f:M13{t:J5/4}>

__1566__; [Cuba]; Venice, Paolo Forlani; TOOLEY 84, NL 4F81.

246. Ladder in shield under six-pointed star D; with three rungs (14 x 35 mm); shield 32 x 75 mm; total height 97 mm. N18<f:M13{t:J5/4}>

1566; [Palestine]; Venice, Paolo Forlani; TOOLEY 193, NL 2F227

1566; [Egypt]; Venice, Paolo Forlani; n.i. TOOLEY, NL 4F407.

__1568__; [Padua]; Venice, Girolamo Olgiati (Ferrando Bertelli); TOOLEY 429, NL 4F311.

245

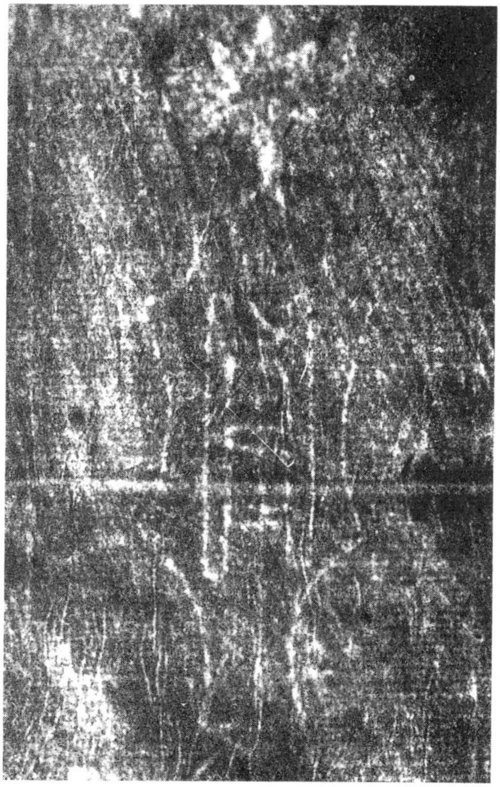

246

247. Ladder in shield under six-pointed star E; with three rungs (14 x 37 mm; one pole ends 1 mm below the other); shield 32 x 76 mm; total height 98 mm. N18<f:M13{t:J5/4}>

1567; *[Piedmont]; Venice, Paolo Forlani; Tooley 451, NL 2F163.*

1568; [Rome]; Venice, Leon; n.i. Tooley, KN 051-21-#6c.

1570; [Rhodes]; n.pl., n.pr.; n.i. Tooley, NL 2F55.

1570 ca; [Morea]; n.pl., n.pr.; n.i. Tooley, NL 2F195.

n.d. [Jerusalem]; n.pl., Claudio Duchetti; Tooley 339, NL 4F115.

248. Ladder in shield under six-pointed star F; with three rungs (13 x 35 mm); shield 30 x 79 mm, pointed base; total height 100mm. N18<f:M13{t:J5/4}>. Other manuals: Heawood 1929 1

1562; *[Germany]; n.pl., Ferrando Bertelli; Tooley 249, NL 2F87.*

1568; [Poland, first of two separate sheets]; Venice, Paolo Forlani; Tooley 454, Doria Atlas 37.

1568; [Gulf of Venice]; Venice, Paolo Forlani; Tooley 587, Doria Atlas 85.

1569 ca; Genova [anonymous view with 25 places identified]; n.pl., n.pr.; n.i. Tooley, CSUF 1328.

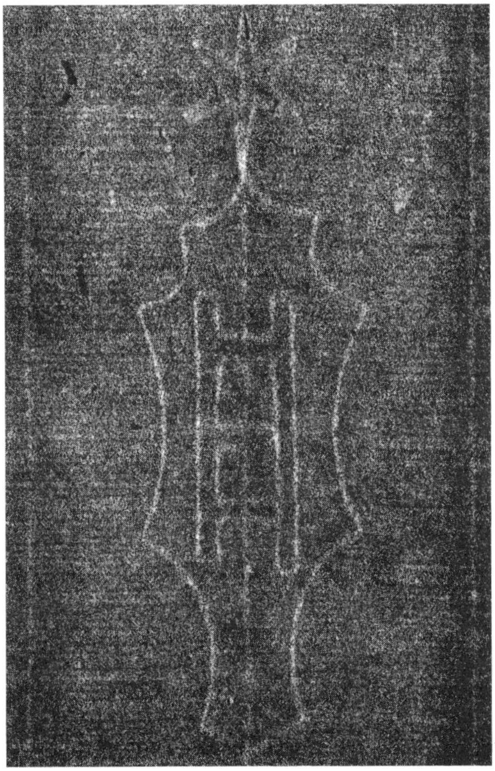

247

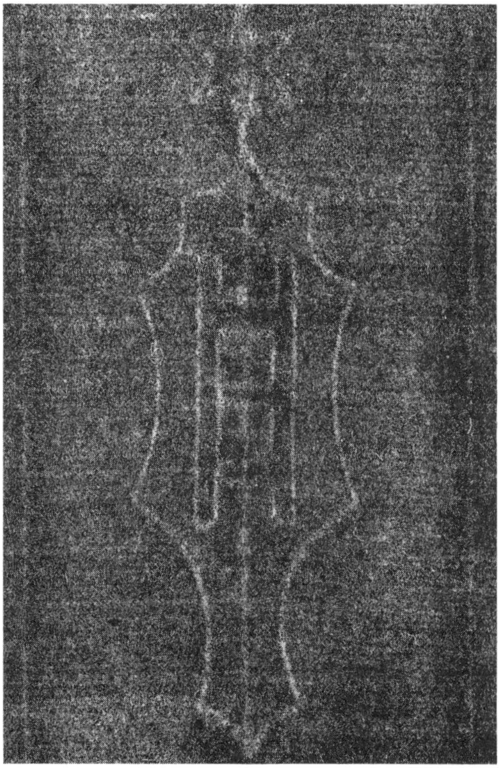

248

249. Ladder in shield under six-pointed star G; with three rungs (9 x 31 mm); shield 31 x 77 mm, flat base; total height 101mm. N18<f:M13{t:J5/4}>

1552; [Germany]; Venice, Giacomo Gastaldi (Gabriel Giolito); Tooley 250, NL 4F168.

1562; [Africa]; Rome, Antonio Lafreri (Venice, Paolo Forlani orig.); Tooley 67, NC 2:130:53 right.

1569; *[Istria]; n.pl., Ferrando Bertelli; Tooley 321, NL 2F155.*

250. Ladder in shield under six-pointed star H; with three rungs (13 x 37; one pole ends 2 mm below other); shield 31 x 74, flat base; total height 96 mm. N18<f:M13{t:J5/4}>

1557; [Vicovaro]; n.pl., Sebastiano di Re; Tooley 594, GHB 234.

n.d. *[Civitella]; n.pl., n.pr.; Tooley 154, NL 2F34.*

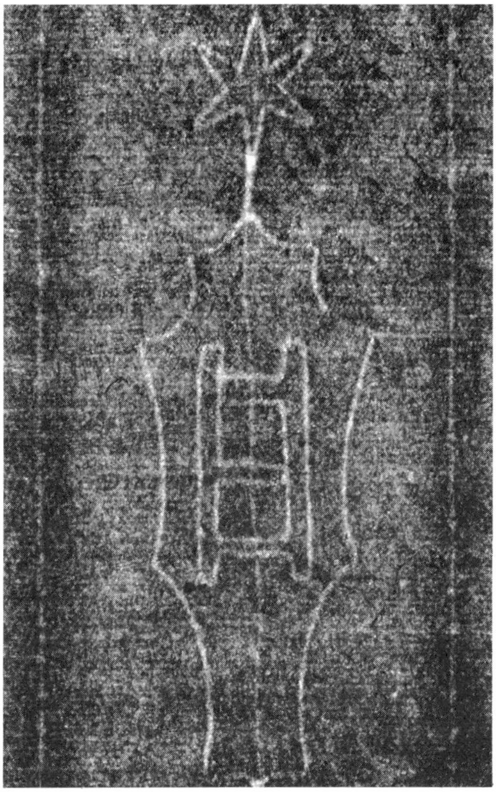

249

250

251. Ladder in shield under six-pointed star I; with three rungs (13 x 36 mm); shield 30 x 81, pointed base; total height 102 mm. N18<f:M13{t:J5/4}>

1566; Venice, Vincenzo Lucchino; TOOLEY 538, GHB 159r.
1570; [Morea]; n.pl., Claudio Duchetti; TOOLEY 400, NL 2F193.
n.d. *[Siena]; n.pl., Claudio Duchetti; TOOLEY 522, NL 2F167.*

252. Ladder in shield under six-pointed star J; with three rungs (14 x 38 mm); shield 30 x 76 mm; total height 97 mm. N18<f:M13{t:J5/4}>. Other manuals: BEANS 31 (sim.)

1557; *[Vicovaro]; n.pl., Sebastiano di Re; TOOLEY 594, KN 87-GR-13.*

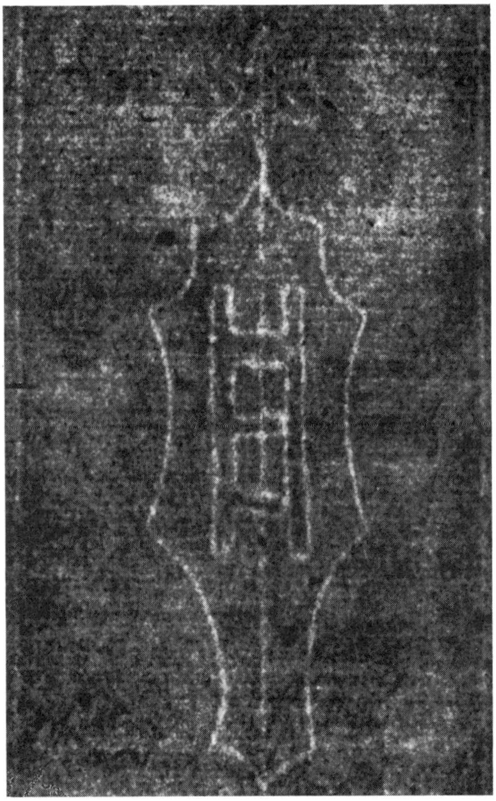

251

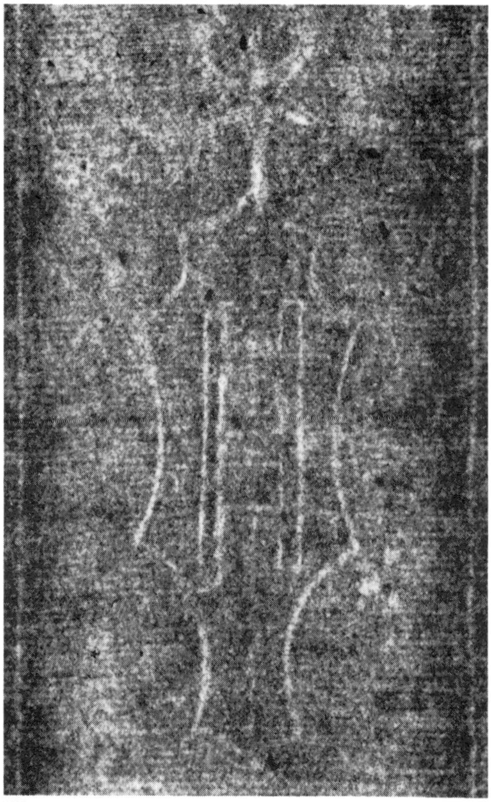

252

253. Ladder in shield under six-pointed star K; with three rungs (14 x 36 mm; one pole crushed at top); shield 30 x 79 mm, pointed base; total height 100mm. N18<f:M13{t:J5/4}>

1555; [Rome]; n.pl., Antonio Lafreri; n.i. Tooley, NL 4F247.

1561; [Second Part of Asia]; Rome, Antonio Lafreri (Venice, Giacomo Gastaldi orig.); Tooley 54, NC 2:130:61.

1561; [Second or third map of Asia series]; Venice, Giacomo Gastaldi; Tooley 54 or 61, Doria Atlas 116.

1565; [Malta, end of first phase of siege of St. Elmo]; Venice, Nicolò Nelli; n.i. Tooley, Doria Atlas 81.

n.d. *[Tiber]; n.pl., n.pr.; n.i. Tooley, NL 2F124.*

n.d. [W. Dalmatia]; Venice, Paolo Forlani (Giacomo Gastaldi); Tooley 29 (w. sheet), CSUF 65277.

254. Ladder in shield under six-pointed star L; with three rungs (14 x 38 mm; one pole ends 1mm below other); shield 33 x 82 mm, very pointed base; total height 114mm. N18<f:M13{t:J5/4}

1562; [Africa, 2 sheets]; Venice, Paolo Forlani; Tooley 67, Doria Atlas 99(i).

1570; *[Cyprus]; Venice, Paolo Forlani; Tooley 185, NL 2F223.*

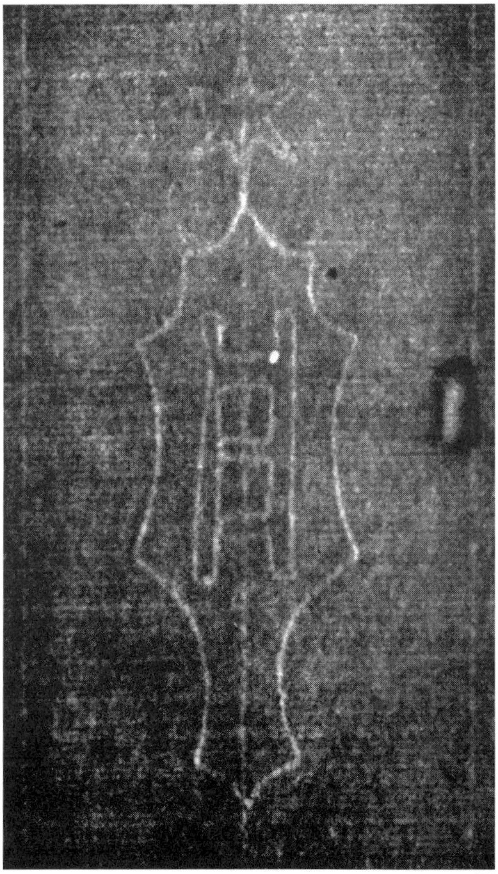

253

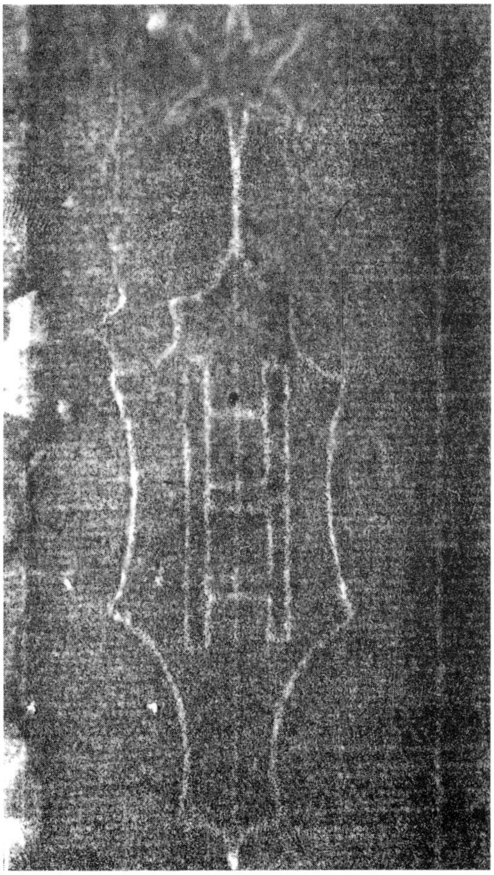

254

255. Ladder in shield under cross pommy A; with three rungs (16 x 35 mm); shield 31 x 80, narrow base; total height 108 mm. N18<f:M13{t:S1}>. Other manuals: Beans 32 (sim.)

1559; [Spain]; Rome, Vincenzo Lucchino; Tooley 530, NL 4F189v.

1561; [Third Part of Asia]; Rome, Antonio Lafreri (Venice, Giacomo Gastaldi orig.); Tooley 61, NC 2:130:62 left.

1561; [Third Part of Asia]; Venice, Giacomo Gastaldi; Tooley 61, Doria Atlas 117(i).

1561; [Third Part of Asia]; Venice, Giacomo Gastaldi; Tooley 61, Doria Atlas 117(ii).

1561; [Third Part of Asia]; Venice, Giacomo Gastaldi; Tooley 61, Doria Atlas 118.

1566; [Frisia]; Rome, Jacob von Daventer; Tooley 221, GHB 220.

n.d. *[Ancona]; n.pl., n.pr.; Tooley 106, CSUF 1321r.*

n.d. [Ancona]; n.pl., n.pr.; Tooley 106, CSUF 1321l.

256. Ladder in shield under cross pommy B; with three rungs (16 x 36 mm); shield 33 x 75, wide base; total height 102 mm. N18<f:M13{t:S1}>

1558; [Great Britain]; Rome, Sebastiano di Re; Tooley 271, CSUF ac 1071.

1559; [West Asia]; Venice, Fabio Licinio (Giacomo Gastaldi); Tooley 46, GHB 207v.

1559; [West Asia]; Venice, Fabio Licinio (Giacomo Gastaldi); Tooley 46, GHB 207r.

1559; [Spain]; Rome, Vincenzo Lucchino; Tooley 530, NL 4F189r.

1560 ca; [Greece]; Venice, Giacomo Gastaldi; Tooley 280, NL 4F354.

1561; [Third Part of Asia]; Rome, Antonio Lafreri (Venice, Giacomo Gastaldi orig.); Tooley 61, NC 2:130:62 right.

1564*; [Tuscany]; Rome, n.pr.; Tooley 568, NL 4F336.*

1566; [Malta]; Rome, Antonio Lafreri; Tooley 361, NL 2F174.

1570; [East Mediterranean]; Rome, Antonio Lafreri; Tooley 184, GHB 223v.

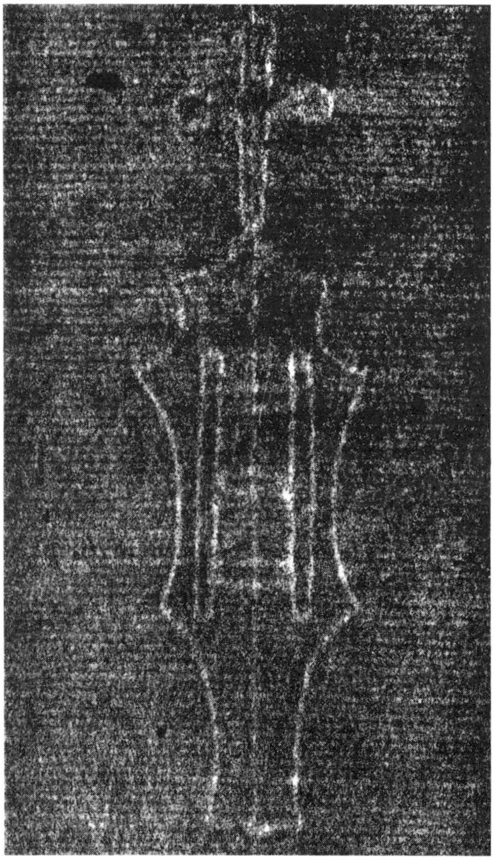

255

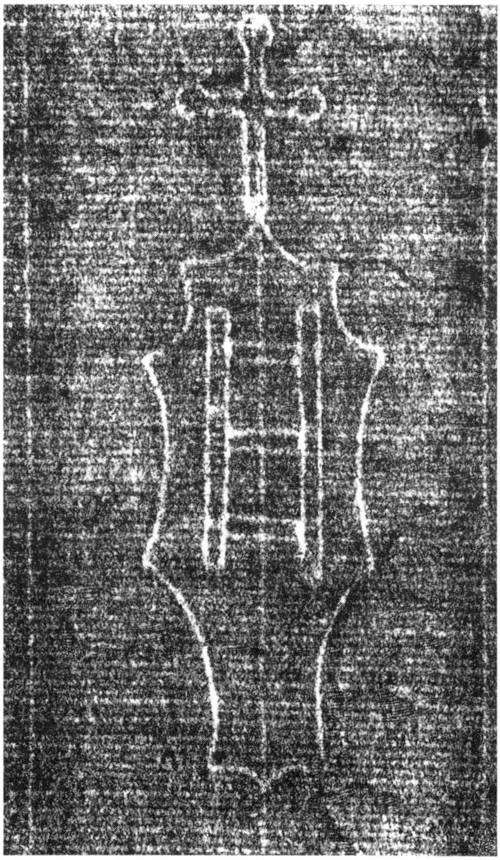

256

257. Ladder in double-outline shield-cum-cross; with three rungs (15 x 27); height of interior line of shield 54 mm; height of exterior line, including cross, 88mm. N18<f:M13{t:S1}>. Other manuals: BEANS 9 (sim.); BRIQUET 5930 (sim.)

1540 ca; [Rome]; n.pl., Enea Vico (Antonio Salamanca); n.i. TOOLEY, NL 2F133.

258. Horn in shield; shield 45 x 47 mm touching both sides of horn; also within shield is six-pointed star. O3/2-J5/4<f:M13>. Other manuals: BRIQUET 7862 (sim.)

1590 ca; Palmae civitas in Patria Foro Juliensis [Palmanova] 365 x 475 mm; n.pl., n.pr.; n.i. TOOLEY, NL 2F126.

257

258

259

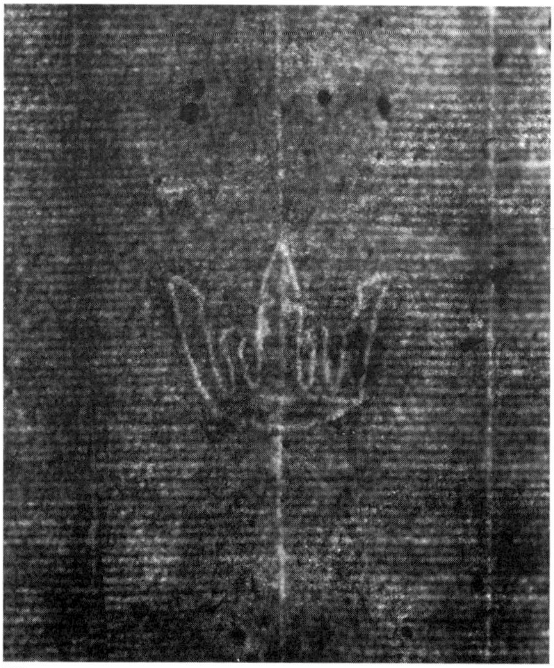

260

259. Crown with four points, under a trefoil; even triangular points in profile: base 28 mm, spread 42 mm, height 31 mm. R3<t:G4>

NB. 'base' figure is the horizontal width of the base; 'spread' is the distance (not necessarily horizontal) between the tips of the left and right points; 'height' is distance along axial chain line between uppermost and lowermost wires of the watermark. Crowns 'in profile' have a single line at bottom.

1571; [Turkish Navy]; n.pl., Ferrando Bertelli; TOOLEY 609, NL 4F108r.

1571; *[Turkish Navy]; n.pl., Ferrando Bertelli; TOOLEY 609, NL 4F108v.*

260. Crown with five points A; crude, uneven crown in profile: base 21, spread 30, height 24. R3

1558; *[Tuscany]; n.pl., Girolamo Bel'Armato; TOOLEY 570, NL 4F334.*

n.d. [Sopoto (Albania)]; n.pl., Antonio Lafreri; TOOLEY 525, NL 2F54.

261. Crown with five points B; even points: base 25, spread 36, height 22. R3

1572; *[Italy]; Venice, Domenico Zenoi;* TOOLEY *334, NL 4F205.*

262. Crown with five points C; similar to previous (#261) but larger; base 33, spread 50, height 31. R3. Other manuals: HEAWOOD CAT 997 (sim.)

1570 ca; [Venice]; n.pl., Donatum Bertelli; n.i. TOOLEY, NL 4F285.

261

262

263

264

263. Crown with five points under a star with six points A; uneven triangular points: base 29, spread 31, height 53, star 22. R3{t:J5/4}

n.d. *[Lake Geneva]; n.pl., n.pr.; Tooley 238, CSUF DDSW 65277.*

264. Crown with five points under a star with six points B; uneven triangular points: base 26, spread 27, height 51, star 20; sim to #269. R3{t:J5/4}

1580 ca; [Rome, Antiqua]; Rome, Ambrogio Brambilla; n.i. Tooley, NL 4F253.

n.d. *[Mirandola]; n.pl., Henricus van Schoel; Tooley 397, NL 4F111.*

265. Crown with five points under a star with six points C; uneven points; 2nd and 4th are stalks: base 28, spread 51, height 75, star 26. R3{t:J5/4}

1564; *[Italy]; Venice, Giacomo Gastaldi; Tooley 331, NL 2F110.*

1564; [Italy (index)]; Venice, Giacomo Gastaldi; Tooley 331, NL 2F108.

1564; [Asia (index)]; Venice, Giacomo Gastaldi; Tooley 59, NL 2F234(i).

266. Crown with five points under a star with six points D; uneven points; 2nd and 4th are stalks: base 29, spread 54, height 80, star 23; sim to #271. R3{t:J5/4}

1564; [Asia]; Venice, Giacomo Gastaldi; Tooley 52, NL 2F219.

1564; [Asia (index)]; Venice, Giacomo Gastaldi; Tooley 59, NL 2F234(ii).

n.d. *[Belgium]; n.pl., n.pr.; Tooley 136, NL 2F82.*

265

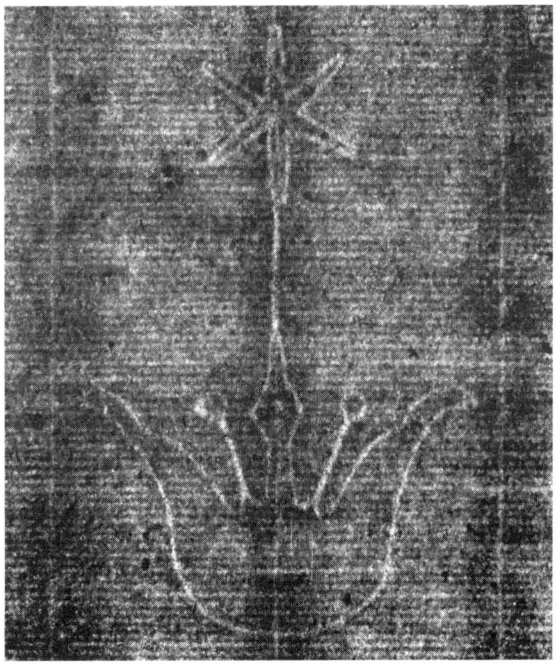

266

267

268

267. Crown with five points under a star with six points E; uneven points; 2nd and 4th are stalks: base 29, spread 54, height 81, star 24; sim to #270. R3{t:J5/4} Other manuals: BEANS 14 (sim.)

1561; [First or second map of Asia series]; Venice, Giacomo Gastaldi; TOOLEY 48 or 54, Doria Atlas 115.

1564; *[Asia (Index)]; Venice, Giacomo Gastaldi; TOOLEY 331, NL 2F109.*

268. Crown with five points under a star with six points F; uneven points; 3rd standing tall in form of trefoil; in profile: base 28 (curved), spread 47, height 64, star 33. R3{t:J5/4}. Other manuals: BRIQUET 4833 (sim.)

1570; *[Turkish army]; n.pl., n.pr.; TOOLEY 609, NL 4F105.*

269. Crown with five points under a star with six points G; uneven points; 2nd and 4th very small: base 26, spread 40, height 60, star 21. R3{t:J5/4}

1557; *[Naples, Kingdom]; Venice, n.pr.; TOOLEY 401; NL 4F337.*

1559; [Hungary]; Rome, Antonio Lafreri; TOOLEY 307, NC 2:130:28.

1566; [Turkish army]; Rome, Antonio Lafreri; TOOLEY 610, NL 2F48.

270. Crown with five points under a star with six points H; uneven points; 2nd and 4th very small: base 23, spread 32, height 66, star 24. R3{t:J5/4}

1569; *[Holy Fathers of Trento]; Venice, n.pr.; n.i. TOOLEY, NL 2F248.*

269

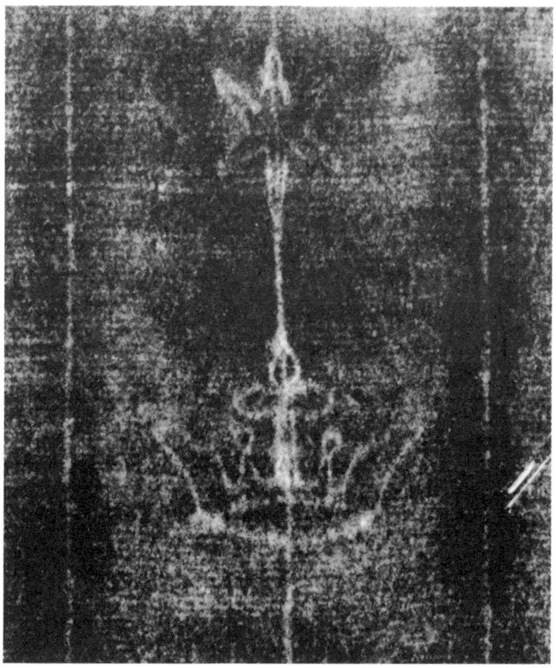

270

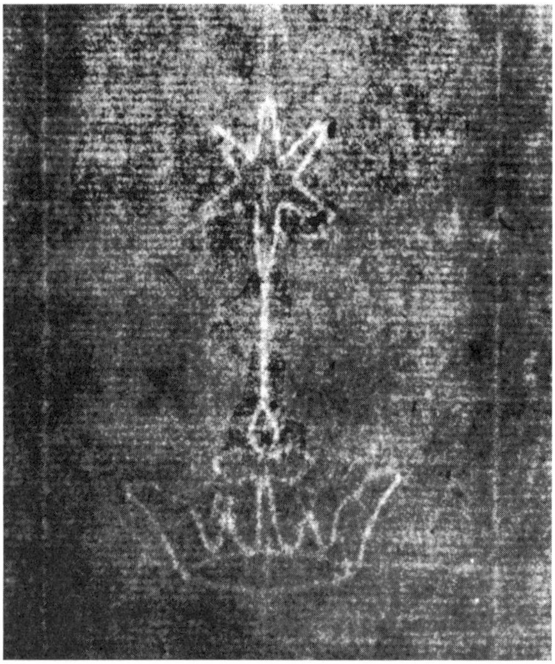

271

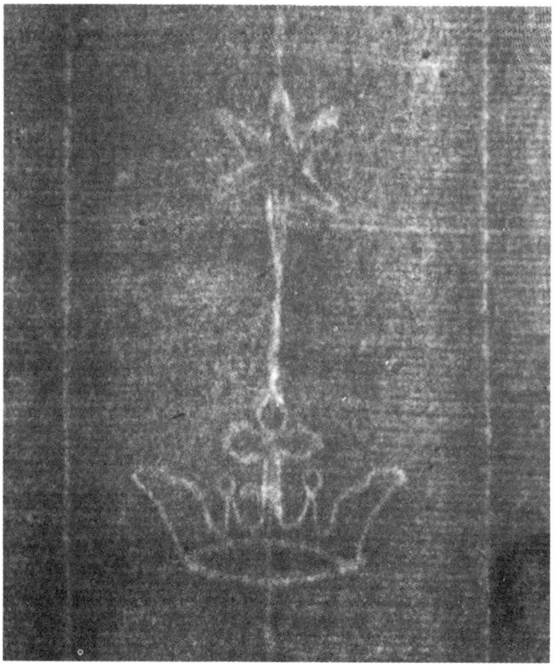

272

271. Crown with five points under a star with six points I; uneven points; 2nd and 4th very small: base 23, spread 38, height 65, star 25; sim to #274. R3{t:J5/4}

__1566__; [Malta]; Rome, Antonio Lafreri; Tooley 362, NL 2F182.

272. Crown with five points under a star with six points J; uneven points; base 24, spread 38, height 67, star 25; sim to #275. R3{t:J5/4}

__1575__; Le sette chiese di Roma; Rome, Antonio Lafreri; n.i. Tooley, NL 4F252.

273. Crown with five points under a star with six points K; uneven points: base 22, spread 37, height 51, star 20. R3{t:J5/4}. Other manuals: BRIQUET 4835

1550 ca; [Boulogne]; n.pl., n.pr.; TOOLEY 140, KN 4-RHE-14.

274. Crown with five points under a star with six points L; uneven points; base 24, spread 37, height 58, star 25. R3{t:J5/4} Other manuals: BEANS 13 (sim.)

1561; [Rome]; Rome, Sebastianus Clodiensis; TOOLEY 488, NL 4F250.

273

274

275

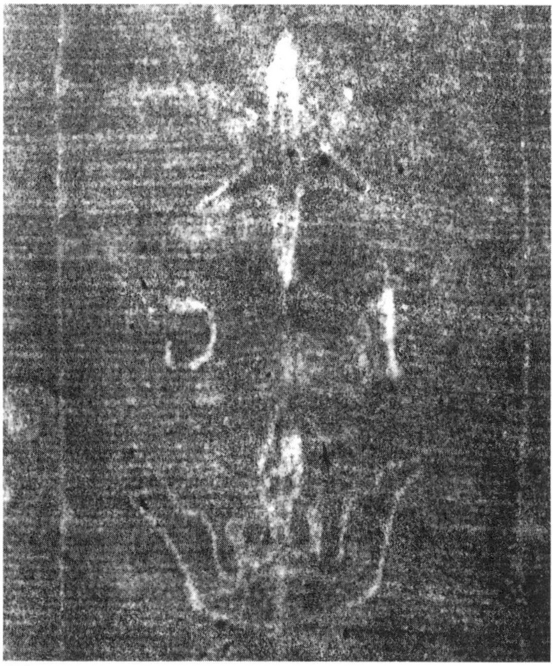

276

275. Crown with five points under a star with six points M; uneven points; 2nd and 4th are stalks; curved base: ?, spread 44, height 58, star 22. R3{t:J5/4}. Other manuals: PICCARD I:II: 106 (sim.)

1563; *[France]; Venice, Giovanni Francesco Camocio; TOOLEY 216, NL 2F76.*

1566; [Friesland]; Venice, Giovanni Francesco Camocio; TOOLEY 218 or 219 (see comment in Sotheby's), Doria Atlas DCG 31.

276. Crown with five points under star with "B" and "G"; uneven points, 2nd and 4th very small; base 25, spread 41, height 77, star 33; single-wire letters between crown and star: 'B' to left of axis, 'G' to right. R3-(i:X"BG"){t:J5/4}

1592; [Lafreri-de Nobilibus Atlas]; Boswell front endpaper.

277. Crown with five points under fleur-de-lys; uneven points; base 21, spread 34, height 22, star 44. R3{t:G6/2}

1585; [Naples]; Rome, Claudio Duchetti; Tooley 412, NL 4F233.

***1586**; [Pozzuoli]; Rome, Claudio Duchetti (Ambrogio Brambilla); n.i. Tooley, NL 4F346.*

278. Crown with five points in circle over "AF" A; curved baseline; circle 46 x 47 mm; below, in double outline, are letters 'AF'. R3<f:U1>{b:(i:X"AF")}. Other manuals: Beans 71 (sim.)

1585 ca; [Shetland Islands]; n.pl., n.pr.; Tooley 197, Boswell 207r.

***1585** ca; [Madagascar]; n.pl., n.pr.; Tooley 503, Boswell 223r.*

1585 ca; [Sri Lanka]; n.pl., n.pr.; Tooley 547, Boswell 224r.

277

278

279

280

279. Crown with five points in circle over "AF" B; curved baseline; circle 47 x 48 mm; A's cross-stroke is flat; F not hunched over. R3 <f:U1> {b:(i:X"AF")}

n.d. *[Algiers]; Rome, Mario Cartaro; n.i. Tooley; GHB 265.*

280. Crown with five points in shield A; uneven points; crown base 18 mm, spread 24 mm; inside a 9-sided shield, 50 mm tall, 30 mm at widest; with crescent (8 mm dia, opening upwards) below crown. R3<f:M13>. Other manuals: Heawood Cat 1127

1620; *[Frascati] 365 x 1455 mm; Rome, Matteo Greuter; n.i. Tooley, NL 2F118.*

281. Crown with five points in shield B; uneven points; with crescent as #280 but shield less lopsided. R3<f:M13>. Other manuals: Heawood Cat 1127; Briquet 5326 (sim.)

__1592__; [Catania]; Rome, Nicolas van Aelst (Antonio Stizza); n.i. Tooley, NL 4F217.

282. Crown with seven points; even triangular points; below are single-wire letters 'TB'; at bottom letter 'P'; 32, 46, 81. R3{b:(i:X"TB")-{b:(i:X"P")}. Other manuals: Massari (1983) 127a (sim.)

__n.d.__ [Dalmatia]; n.pl., Stefano Scolari; n.i. Tooley, NL 2F206.

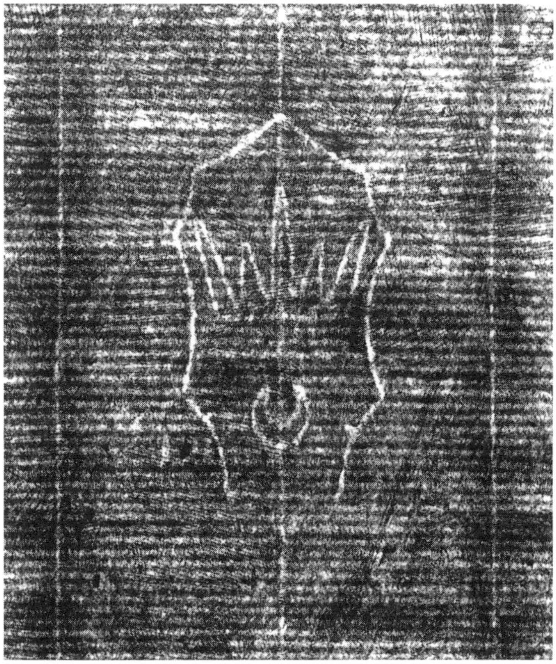

281

282

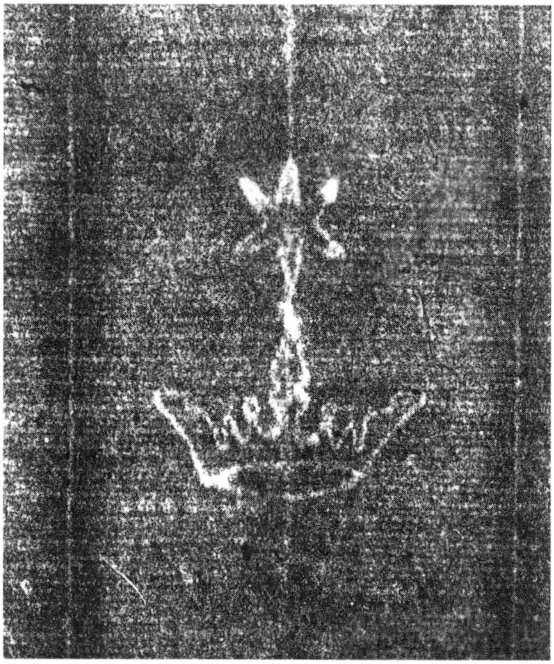

283

284

283. Crown with seven points under six-pointed star A; uneven points: 21, 37, 45, 17. R3{t:J5/4}. Other manuals: PICCARD I:II:113 (sim.)

n.d. *[Majorca/Minorca]; n.pl., Ferrando Bertelli; TOOLEY 355/391, NL VCA 6/7.*

284. Crown with seven points under six-pointed star B; uneven points: 23, 37, 50, 21. R3{t:J5/4}. Other manuals: BEANS 15 (sim.); PICCARD I:II:113 (sim.)

1566*; [Transilvania, Hungary, Romania]; Venice, Paolo Forlani; TOOLEY 29, NL VCA 10.*

285. Crown with seven points under six-pointed star C; uneven points: 3, 34, 48, 20. R3{t:J5/4}. Other manuals: PICCARD I:II:113 (sim.)

1556; *[Piedmont]; Venice, Giacomo Gastaldi (Gabriel Giolito de Ferrari); TOOLEY 448,* NC *1:48:17.*

1559; [SE Europe]; [Venice], Fabio Licinio; TOOLEY 25, NC 1:48:30.

1570; [Germany]; Venice, Giacomo Gastaldi; TOOLEY 257, NL 4F168.

286. Pennant with three tips; pole with ball on top; letters "RA" in single wire, split to either side of pole; 67mm tall. Other manuals: HEAWOOD CAT 1374 (sim.) T2/1{L:(i:X"R")}{r:(i:X"A")}

1580 *ca; [Venice, laguna]; Venice, Ferrando Bertelli; n.i. TOOLEY, NL 4F321.*

285

286

287

288

287. Two circles on stems in the shape of an inverted V (cherries?); circles 22 mm dia, with 'e'; single-wire 'x' below. (2*(U1))[(i:X"e"]-{b:(i:X"x")}

***1590** ca; [Florence]; Siena, Matteo Florimi; n.i.* *TOOLEY, VAN DER HEYDEN 24, NL 2F117.*

288. Lozenge in circle under six-pointed star; all four points touching circle (41 x 43 mm); total height 65mm. U3<f:U1>{t:J5/4}

***n.d.** Ostia. Il vero disegno de tutti li forti che si son fatti dintorno ad Ostia, 290 x 320 mm; n.pl., n.pr.; n.i. TOOLEY, NL 2F123.*

289. Lozenge containing six-pointed star in circle A; circle 41 x 45mm. U3[J5/4]<f:U1>

1544; [Spain]; Venice, Michele Tramezzino (Sebastiano di Re); TOOLEY 527, NL 4F188.

1558; *[N. Europe]; Venice, Michele Tramezzino; TOOLEY 40, NL 4F116.*

290. Lozenge containing six-pointed star in circle B; circle 42 x 45 mm. U3[J5/4]<f:U1>

1558; *[Tuscany]; n.pl., Girolamo Bell'Armato, TOOLEY 570, GHB 200.*

1558; [Belgium, France]; Rome, Pirro Ligorio (orig?); TOOLEY 132, NC 1:48:9.

1559; [Spain]; Rome, Michele Tramezzino (Sebastiano di Re); TOOLEY 529, CSUF ac 1151.

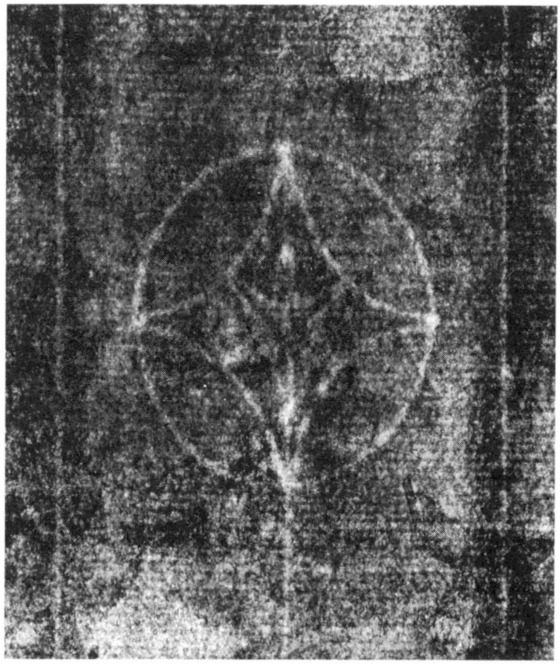

289

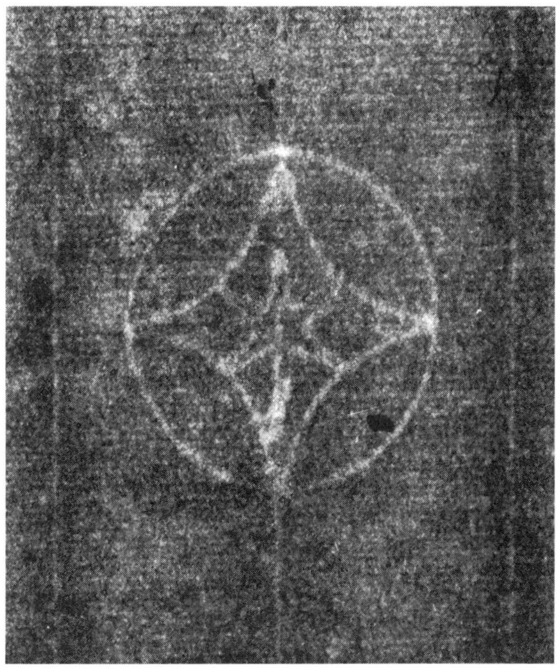

290

291

292

291. Lozenge containing six-pointed star in circle C; circle 42 x 45 mm. U3[J5/4]<f:U1>

1558; [France]; Rome, Pirro Ligorio (Oronce Finé, 1538 orig.); TOOLEY 208, NC 1:48:8. Other manuals: BORRONI SALVADORI 88 (sim.)

1565; *[Italy]; Venice, Ferrando Bertelli; TOOLEY 333, NL 4F203r.*

1565; [Italy]; Venice, Ferrando Bertelli; TOOLEY 333, NL 4F203v.

292. Lozenge containing six-pointed star in circle D; circle 45 x 47 mm. U3[J5/4]<f:U1>. Other manuals: BEANS 46 (sim.); HEAWOOD 1929 16 (sim.); BORRONI SALVADORI 87 (sim.)

1555; *[Switzerland]; Rome, Antonio Salamanca (Jacob Bos); TOOLEY 534, NL 4F161v.*

1555; [Switzerland]; Rome, Antonio Salamanca (Jacob Bos); TOOLEY 534, NL 4F161r.

1559; [Hungary]; Rome, Michele Tramezzino; TOOLEY 308, NL 4F182d (ms).

293. Lozenge containing six-pointed star in circle E; circle 37 x 39 mm. U3[J5/4]<f:U1>

1555; [Switzerland]; Rome, Antonio Salamanca (Jacob Bos); TOOLEY 534, NL 4F162r.

1560; *[Switzerland]; Rome, Jacob Bos; TOOLEY 534, GHB 229v.*

294. Lozenge containing six-pointed star in circle F; circle 37 x 38 mm. U3[J5/4]<f:U1>

1555; *[Switzerland]; Rome, Antonio Salamanca (Jacob Bos); TOOLEY 534, NL 4F162v.*

1560; [Switzerland]; Rome, Jacob Bos; TOOLEY 534, GHB 229r.

293

294

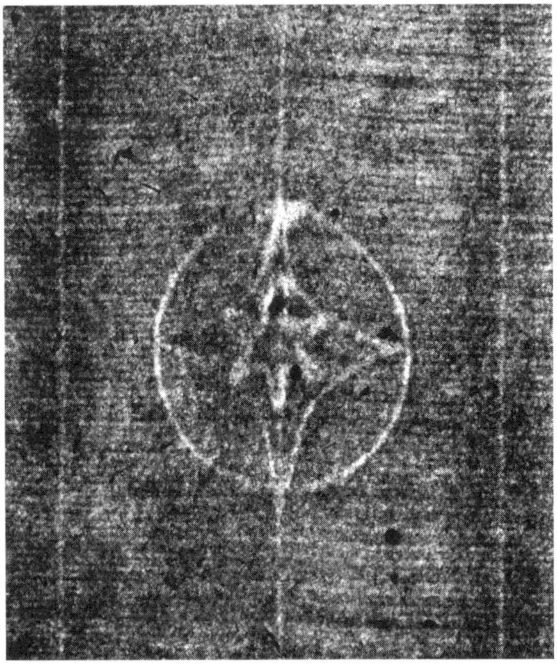

295

296

295. Lozenge containing six-pointed star in circle G; circle 34 x 37 mm. U3[J5/4]<f:U1>

n.d. *[Florence]; n.pl., Claudio Duchetti; TOOLEY 206, NL 4F219.*

296. Lozenge containing six-pointed star in circle H; circle 35 x 36 mm. U3[J5/4]<f:U1>

1570; *[Morea]; n.pl., Claudio Duchetti; TOOLEY 400, Boswell 218r.*

Letters such as the following would normally appear as countermarks to the main watermark on the mould. However, in mid-sixteenth-century Italian maps, cases of such countermarks are rare. The following small letters were found without corresponding watermarks.

297. Letters 'A9' in single wire; 19 mm tall. W1[V2](i:"A9")

1554; *[South Asia]; n.pl., Giacomo Gastaldi (Giovanni Battista Ramusio); n.i.* TOOLEY, *KN 002-RN-6v.*

298. Letters '3C' in single wire under trefoil in circle; above spiral; circle 43 x 52mm. W1[V2](i:"3C")[G4]<f:U1>

1580 *ca; [Hebrew Army]; n.pl., n.pr.; n.i.* TOOLEY, *NL 2F249.*

297

298

299

300

299. Letters 'AB' in single wire under trefoil; total height 31.5 mm. W1 (i:"AB"){t:G4}

1556; [*Piedmont*]; *Venice, Giacomo Gastaldi (Fabio Licinio); Tooley 448, GHB 379.*

300. Letters 'BAB' in single wire under trefoil; 'A' small; total height 47 mm. W1(i:"BAB"){t:G4}

1600 *ca*; [*Reggio Emilia*]; *n.pl., Justus Sadeler; n.i. Tooley, NL 2F128.*

301. Letters 'GA' in single wire under trefoil; total height 38 mm. W1(i: "GA"){t:G4}

1599; [*Italy*]; *Venice, Giacomo Gastaldi; n.i.* TOOLEY, *NL 2F106r.*

302. 'NB' in single wire. W1(i:"NB")

1566; [*Africa*]; *Venice, Paolo Forlani;* TOOLEY *70, NL 4F394.*

301

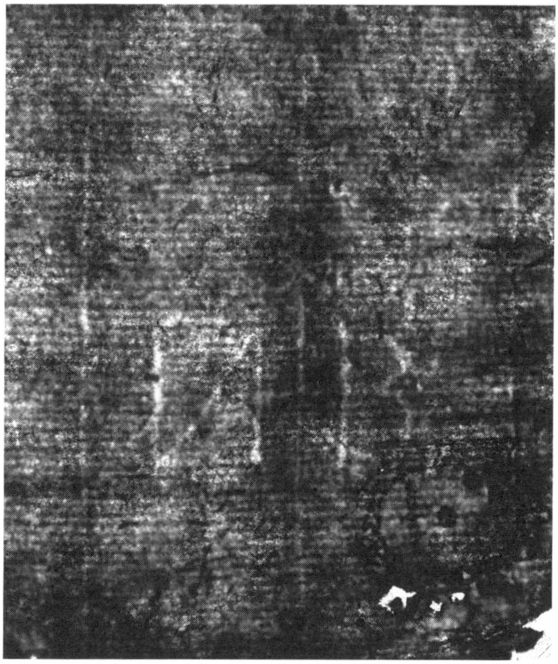

302

303

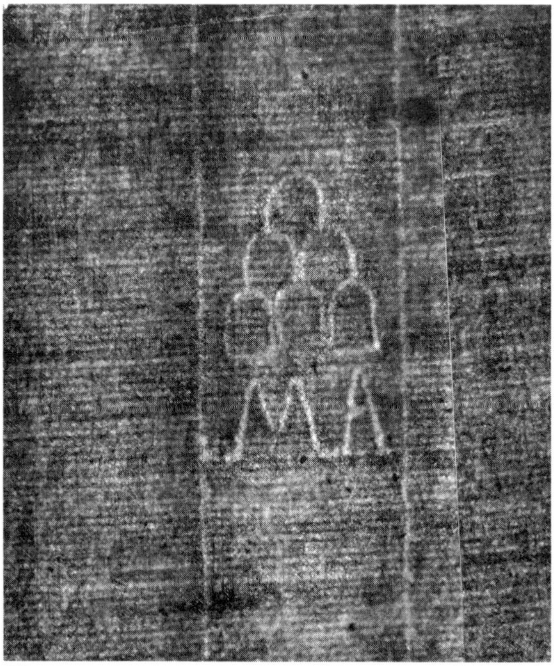

304

303. 'BS' under trefoil; total 26 mm. W1(i:"BS"){t:G4}

1538; [Cyprus]; Venice, Matteo Pagano; Tooley 179, NC 1:48:26 l.

304. 'AM' or 'MA' in single wire under nine hills; dimensions 21 x 38 mm. W1(i:"AM"){t:(9*(J3))}

1600; [Pavia]; n.pl., Johannes Maius Romanus; n.i. Tooley, NL 4F241.

305. Letters 'BC' in double wire in oval A, separated by a heart in single wire, oval 55 x 13 mm. W1[p09]{L:(i:"B")}{r:(i:"C")}

1480; *Venetia. Civitas Venetiarum; [from Breydenbach]; n.i.* T*OOLEY*, *NL 2F145(i).*

1480; Venetia. Civitas Venetiarum; [from Breydenbach]; n.i. TOOLEY, NL 2F145(ii).

306. Letters 'BC' in double wire in oval B, separated by a heart in single wire, oval 56 x 14 mm. W1[p09]{L:(i:"B")}{r:(i:"C")}

1480; *Venetia. Civitas Venetiarum; [from Breydenbach]; n.i.* T*OOLEY*, *NL 2F145(iir).*

1480; Venetia. Civitas Venetiarum; [from Breydenbach]; n.i. TOOLEY, NL 2F145(iic).

305

306

307

308

307. Letters 'GA' in single wire in heart (37 x 40 mm). W1(i:"GA")<f:p09>

1590 *ca; [Parma]; Siena, Matteo Florimi (Marcus Clodianus);* VAN DER HEYDEN *35, KN 08-ANX-15.*

308. Letters 'PF' in single wire in heart (36 x 43 mm). W1(i:"PF") <f:p09>

1590 *ca; [Moscow]; Siena, Matteo Florimi; n.i.* TOOLEY, VAN DER HEYDEN *31, NL 2F213.*

309. Letters 'IHS' in double wire; crossbar of H extends to a cross; all in circle (45 x 48 mm, somewhat crooked). W1[S1] (i:"IHS")<f:U1>

***1585** ca; [Piacenza]; n.pl., Matteo Florimi; n.i.* TOOLEY, VAN DER HEYDEN *39, CSUF 1325.*

310. Letter 'M' under fleur-de-lys in shield A; 'M' in double wire, crossed with a bar in single wire; shield 50 x 70 mm. W1(i:"M"){t:G6/2)<f: M13>

***1569**; [Austria]; Venice, Bolognino Zaltieri;* TOO-LEY *126, NL 2F203.*

309

310

311

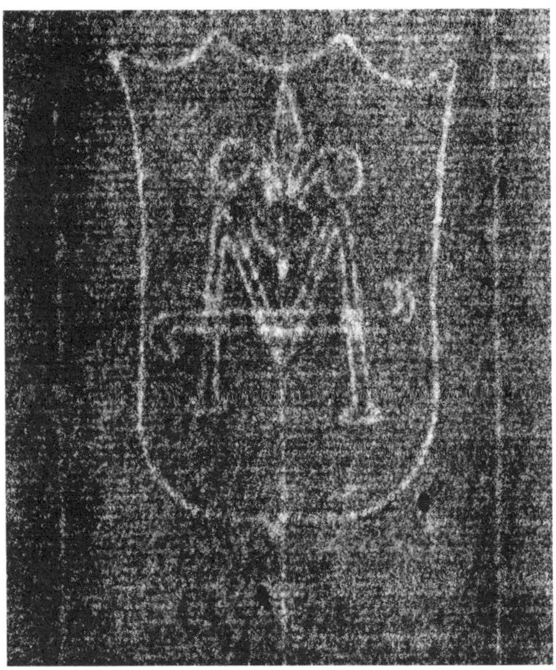

312

311. Letter 'M' under fleur-de-lys in shield B; 'M' in double wire, crossed with a bar in double wire; beneath fleur-de-lys; all in shield 49 x 69mm. W1(i:"M"){t:G6/2)<f:M13>

__1570__; Disegno de l'ordine di tutta l'armata Cristiana congionta insieme al porto di Suda in Candia; Rome, Mario Cartaro; n.i. TOOLEY, *NL 4F102.*

312. Letter 'M' under fleur-de-lys in shield C; 'M' in double wire, crossed with a bar in double wire; beneath a fleur-de-lys; all in a shield 47 x 67 mm. W1(i:"M"){t:G6/2)<f:M13>

1561; [Italy]; Venice, Giacomo Gastaldi; TOOLEY 328, NL 4F200v.

__1570__; [Lombardy]; Rome, Antonio Lafreri (Giorgio Tillman; Giacomo Gastaldi); TOOLEY *354, NL 4F309(l).*

313. Letter 'M' under star in shield A; 'M' in double wire with six-pointed star in shield (thin, with 'shoulders'; 34 x 67 mm). W1(i:"M")[J5/4]<f:M13>

n.d. *[Gradisca]; n.pl., Giacomo Franco; n.i.* Tooley, *NL 2F119.*

314. Letter 'M' under star in shield B; 'M' in double wire with six-pointed star in shield with knobs at each shoulder; notch off top of one shoulder (46 x 70 mm). W1(i:"M")[J5/4]<f:M13>

1585*; [Adriatic Sea]; Venice, Paolo Forlani; n.i.* Tooley, *Boswell 279 (notch off top of one shoulder).*

1585 ca; [Iceland]; n.pl., n.pr.; Tooley 309, Boswell 202r.

313

314

315

316

315. Letter 'M' under star in shield C; 'M' in double wire, with one leg crooked; with six-pointed star in shield with knobs at each shoulder (44 x 69 mm). W1(i:"M")[J5/4]<f:M13>. Other manuals: BEANS 42 (sim.); MASSARI (1980) 145a (sim.)

1573; *[Milano]; Rome, Antonio Lafreri; TOOLEY 390, Boswell 235r.*

316. Letter 'M' under star in shield D; 'M' in double wire; with six-pointed star in shield with knobs at each shoulder (45 x 70mm); less curve to sides. W1(i:"M")[J5/4]<f:M13>

1558; *[Tuscany]; Rome, Claudio Duchetti (Girolamo Bell'Armato); TOOLEY 573, Boswell 196r.*

317. Letter 'M' under star in shield E; 'M' in double wire; with six-pointed star in shield with knobs at each shoulder (43 x 72 mm); one shoulder dropped. W1(i:"M")[J5/4]<f:M13>

1571; *[World]; Plantin, Benedict Arias Montanus; n.i.* TOOLEY, *NL 2F8.*

318. Letter 'M' under star in shield F; 'M' in double wire; with six-pointed star in shield with knobs at each shoulder (46 x 72 mm). W1(i:"M")[J5/4]<f:M13>

1580 ca; [Ptolemaic World]; n.pl., Petri de Nobilibus (orig. Arnold Buckinck (?), ca. 1480); n.i. TOOLEY, NL 4F3r.

1580 ca; [Ptolemaic World]; n.pl., Petri de Nobilibus (orig. Arnold Buckinck (?), ca. 1480); n.i. TOOLEY, NL 4F3v.

1573; *[Milano]; Rome, Antonio Lafreri;* TOOLEY *390, NL 4F230.*

317

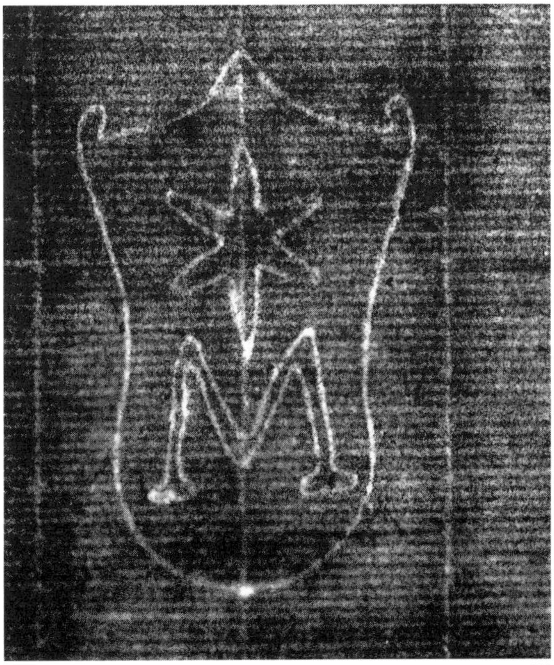

318

319

320

319. Letter 'M' under star in shield G; 'M' in double wire; with six-pointed star in shield with thin waist (44 x 72 mm). W1(i:"M")[J5/4]<f:M13>

***1590** ca; [Pozzuoli eruption 29 Sept. 1538]; n.pl., Petri de Nobilibus; Tooley 458, NL 4F343.*

320. Letter 'M' under star in shield H; 'M' in double wire, one leg crooked; with six-pointed star in shield with knobs at each shoulder (41 x 72 mm). W1(i:"M")[J5/4]<f:M13>

1570 ca; [Siena]; n.pl., Claudio Duchetti; Tooley 522, Boswell 197r.

***1581**; [Genoa]; Rome, Claudio Duchetti; Tooley 243, NL 4F223.*

1590 ca; [Corsica]; n.pl., Fabio Licinio; Tooley 169, Boswell 210r.

n.d. [Tunis]; n.pl., Petri de Nobilibus; Tooley 557 (sans Nobilibus imprint), GHB 401.

321. Letter 'M' under star in shield I; 'M' in double wire; with six-pointed star in shield with knobby shoulders and fairly straight sides (44 x 69 mm). W1(i:"M")[J5/4]<f:M13>

1589; *[Rome, Basilica Santa Maria]; n.pl., Nicolas van Aelst; n.i.* TOOLEY, *NL 4F275.*

322. Letter 'M' under star in shield J; 'M' in double wire; with six-pointed star in shield with pointed shoulders and light curve to sides (37 x 68 mm). W1(i:"M")[J5/4]<f:M13>

1590 *ca; [Campania]; n.pl., Petri de Nobilibus (Gasparo Algaria);* TOOLEY *458, Boswell 227r.*

1590 ca; [Malta]; Rome, Antonio Lafreri (Petri de Nobilibus); TOOLEY 363, Boswell 245r.

321

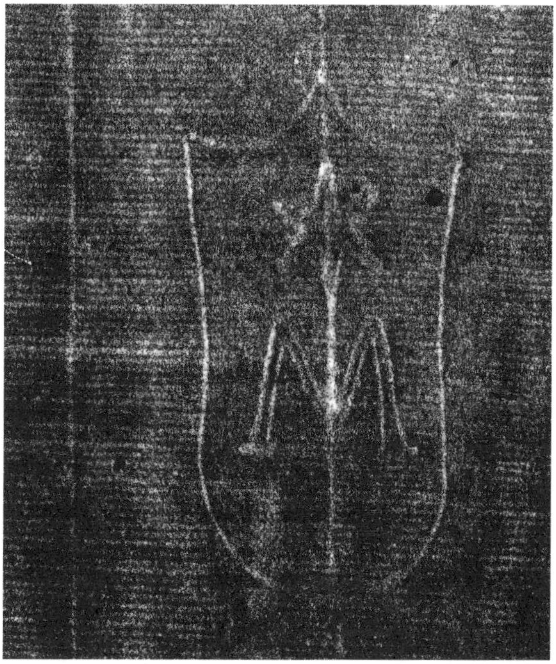

322

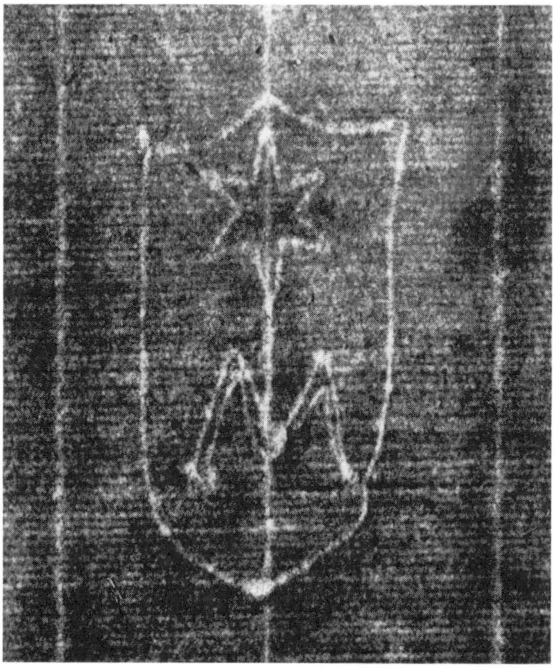

323

324

323. Letter 'M' under star in shield K; 'M' in double wire, one leg ending in point; with six-pointed star in shield with pointed shoulders (36 x 65 mm). W1(i:"M")[J5/4]<f:M13>

1569; *[Ancona]; Ancona, Giacomo Fontana; Tooley 106, NL 4F212v.*

324. Letter 'M' under star in shield L; 'M' in double wire; with six-pointed star in shield with pointed shoulders (38 x 65 mm). W1(i:"M")[J5/4]<f:M13>. Other manuals: Heawood 3024 (sim.)

1569; *[Ancona]; Ancona, Giacomo Fontana; Tooley 106, NL 4F212r.*

325. Letter 'M' under star in shield M; 'M' in double wire; with six-pointed star in shield with pointed shoulders (40 x 63 mm). W1(i:"M")[J5/4] <f:M13>. Other manuals: ZONGHI 1783

1582; [*Ancient Rome*]; *Rome, Claudio Duchetti; TOOLEY 494, KN 051-21-#10.*

326. Letter 'M' under star in shield N; 'M' with long legs in double wire; with six-pointed star in shield with pointed shoulders (37 x 64 mm). W1(i:"M") [J5/4]<f:M13>

1588; [*Pozzuoli*]; *n.pl., Mario Cartaro; n.i. TOOLEY, NL 2F175.*

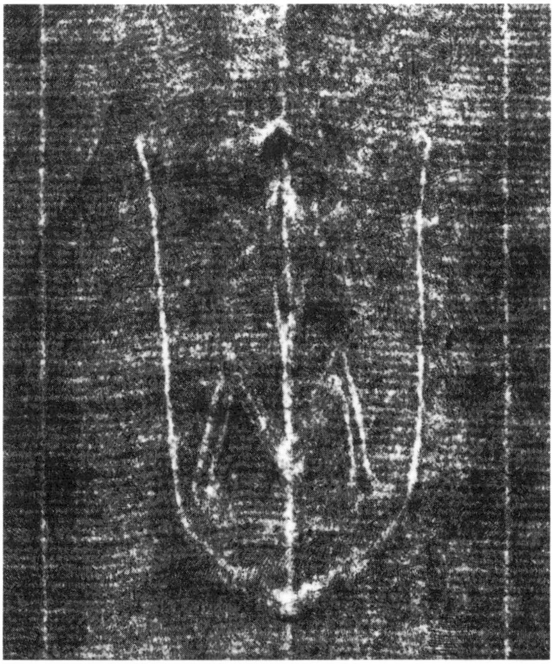

325

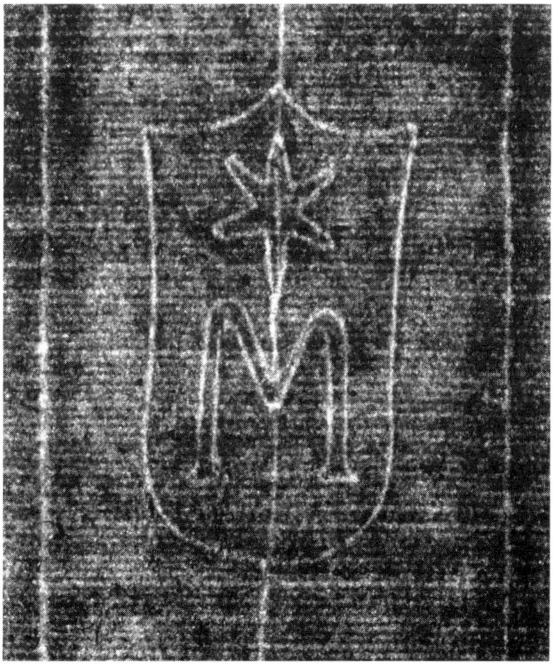

326

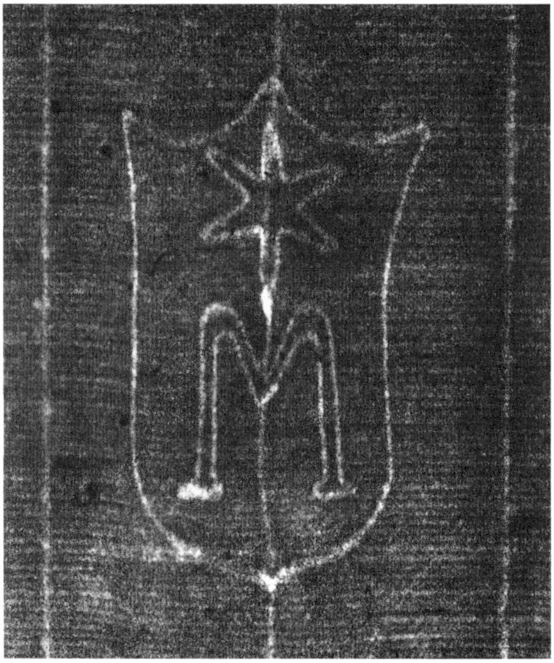

327

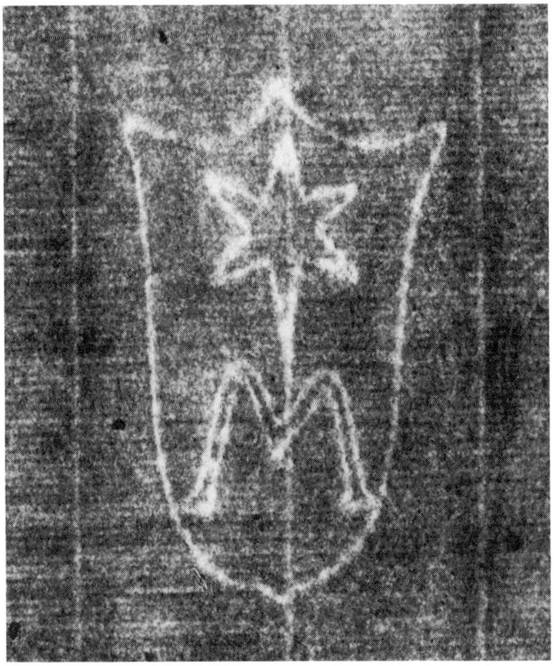

328

327. Letter 'M' under star in shield O; 'M' with long legs in double wire; with six-pointed star in shield with exaggerated pointed shoulders (42 x 68 mm). W1(i:"M")[J5/4]<f:M13>. BEANS 43 (sim.)

***1585** ca; [Parma]; n.pl., Ambrogio Brambilla (Marcus Clodianus); TOOLEY 444, Boswell 237r.*

328. Letter 'M' under star in shield P; 'M' in double wire; with six-pointed star in shield with exaggerated pointed shoulders (44 x 69 mm); axis off central wire. W1(i:"M")[J5/4]<f:M13>

1571; [Europe]; Venice, Paolo Forlani (Claudio Duchetti); TOOLEY 39, GHB 268r.
***1571**; [Europe]; Venice, Paolo Forlani (Claudio Duchetti); TOOLEY 39, GHB 268v.*

329. Letter 'M' under star in shield Q; 'M' in double wire; with six-pointed star in shield with exaggerated shoulders (43 x 66 mm). W1(i:"M")[J5/4] <f:M13>

1570; [*Morea*]; *n.pl., Claudio Duchetti;* TOOLEY *400, NL 2F194.*

330. Letter 'M' under star in shield R; 'M' in double wire; with six-pointed star (one point deformed) in small, blocky shield (40 x 61 mm). W1 (i:"M")[J5/4]<f:M13>

1596; [*Viterbo*]; *n.pl., Tarquinius Ligustri;* TOO-LEY *598, NL 4F294.*

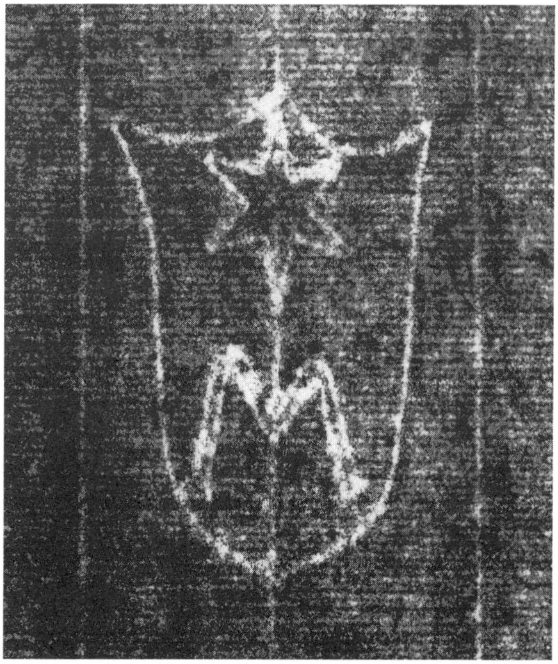

329

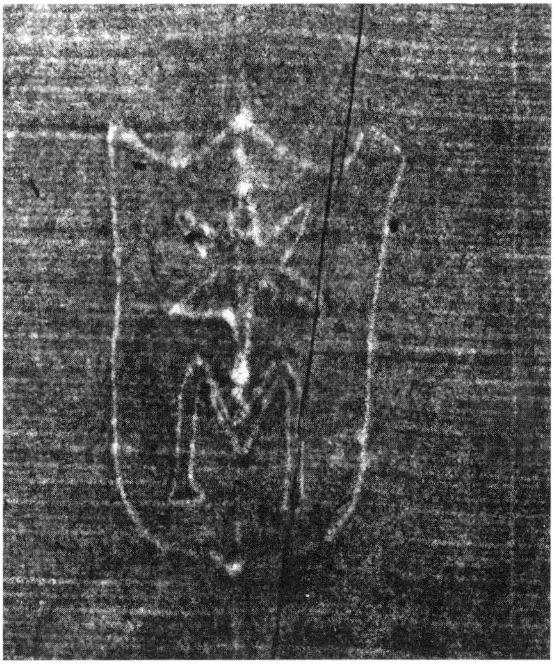

330

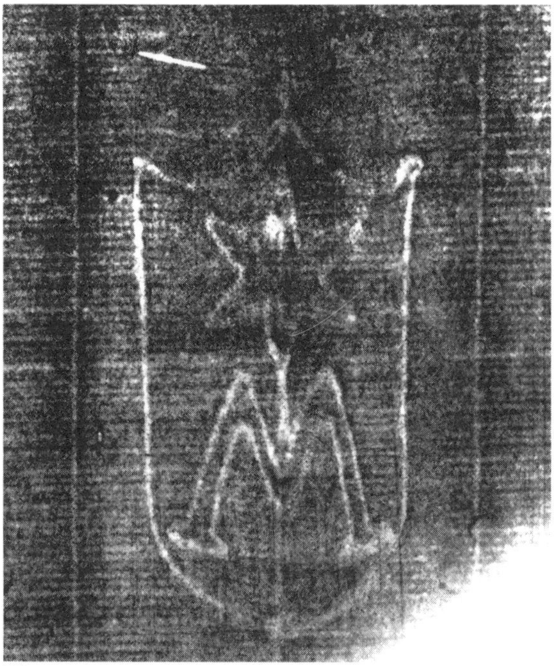

331

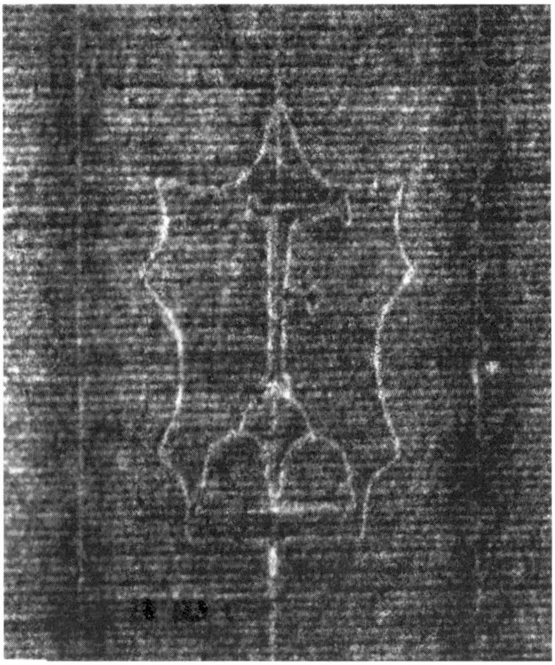

332

331. Letter 'M' under star in shield S; 'M' in double wire; with six-pointed star in shield with straight sides (39 x 68 mm) and one shoulder very sharply pointed. W1(i:"M")[J5/4]<f:M13>

1580 ca; [Rome, Piazza S. Pietro]; n.pl., Nicolas van Aelst; n.i. TOOLEY, NL 4F273.

332. Letter 'F' in double wire over triple-mount A; in shield with sharp corners and points on sides (37 x 58 mm). W1(i:"F")[J5/4]<f:M13>

1602; *[State of Ferrara]; n.pl., Gironimo Dinosso; n.i. TOOLEY, NL 4F300.*

333. Letter 'F' in double wire over triple-mount B; in shield with sharp corners and points on sides (30 x 51 mm). W1(i:"F")[J5/4]<f:M13>

***1580** ca; [Piacenza]; Rome, Henricus van Schoel; n.i.* T*OOLEY*, *NL 4F244.*

334. Letters 'ScC' in shield A; letters in single wire; in rectangular, pointed shield (44 x 35 mm) under trefoil; total 45 mm. W1(i:"ScC")<f:M13>{t:G4}

***1570**; [Central Europe]; Venice, Nicolò Nelli (Andreas Pograbius);* T*OOLEY 501, Boswell 186r.*

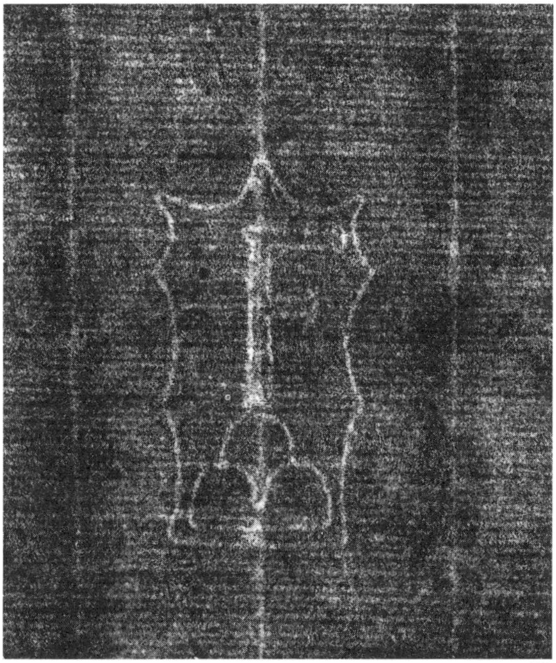

333

334

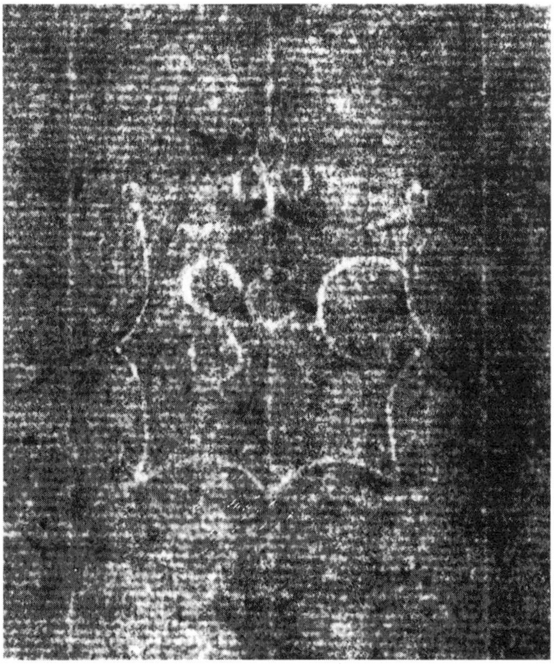

335

335. Letters 'ScC' in shield B; letters in single wire; in rectangular, pointed shield (43 x 37 mm) under trefoil; total 47 mm. W1(i:"ScC")<f:M13>{t:G4}

1570; *[Central Europe]; Venice, Nicolò Nelli (Andreas Pograbius); Tooley 501; Boswell 185v.*

GRAPHIC INDEX

GRAPHIC INDEX

1

2

3

4

5

6

7

8

9

10

11

12

13

14

15

16

17

18

19

20

21

22

23

24

25

26

27

28

29

30

31

32

33

34

35

36

GRAPHIC INDEX

37

38

39

40

41

42

43

44

45

46

47

48

49

50

51

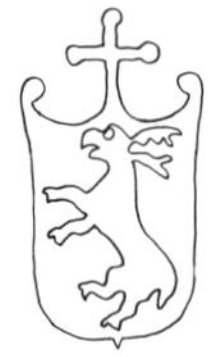
52

53

54

55

56

57

58

59

60

61

62

63

64

65

66

67

68

69

70

71

72

GRAPHIC INDEX

73

74

75

76

77

78

79

80

81

82

83

84

85

86

87

88

89

90

91

92

93

94

95

96

97

98

99

100

101

102

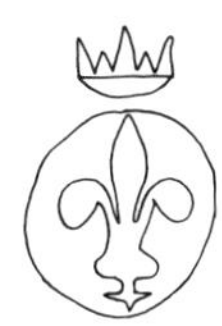
103

104

105

106

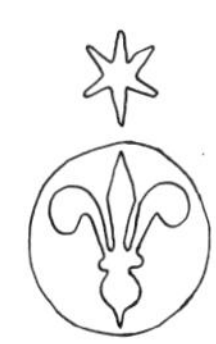
107

108

GRAPHIC INDEX

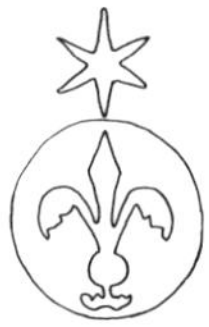
109

110

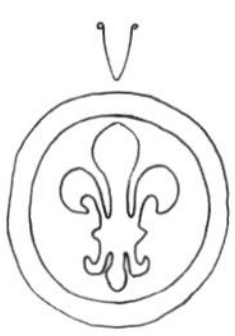
111

112

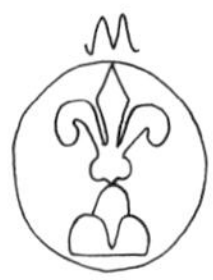
113

114

115

116

117

118

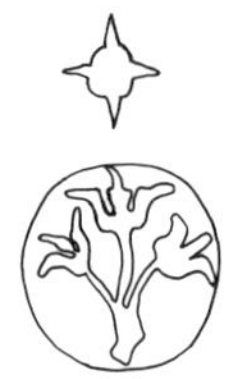
119

120

121

122

123

124

125

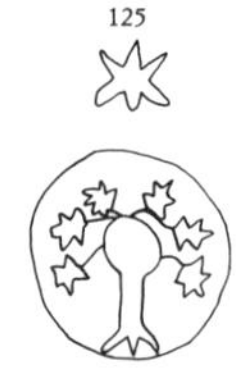
126

127

128

129

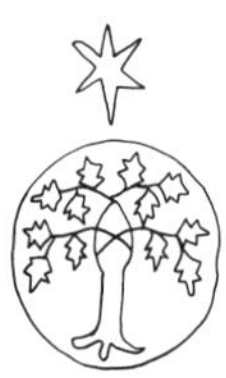
130

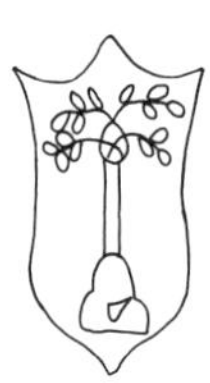
131

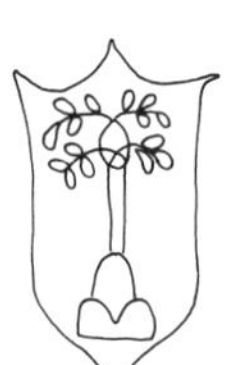
132

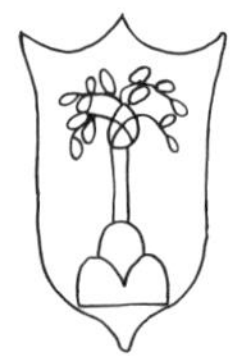
133

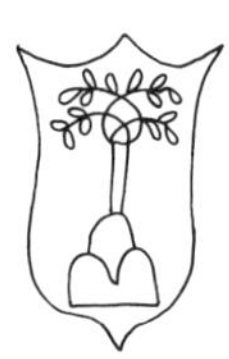
134

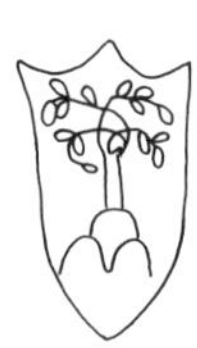
135

136

137

138

139

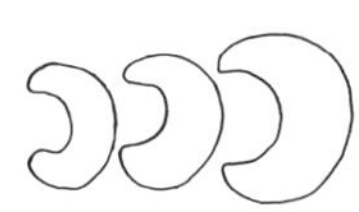
140

141

GRAPHIC INDEX

142

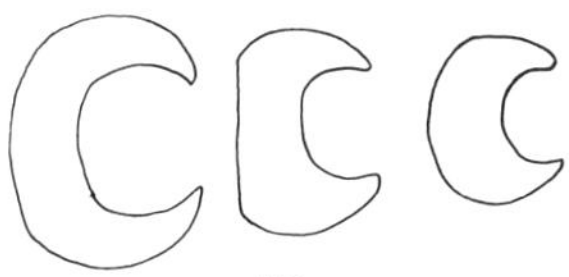
143

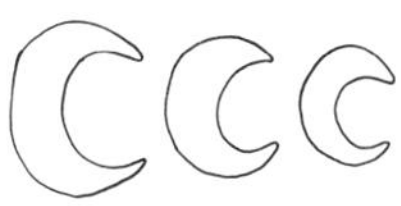
144

145

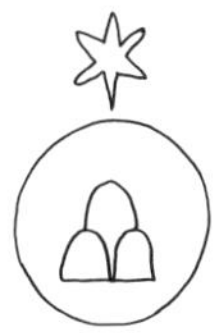
146

147

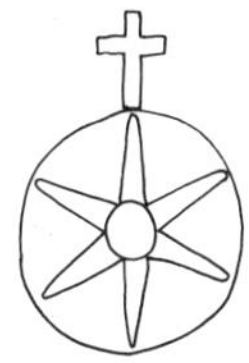
148

149

150

151

152

153

154

155

156

157

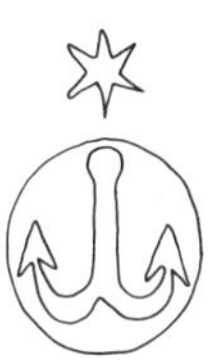
158

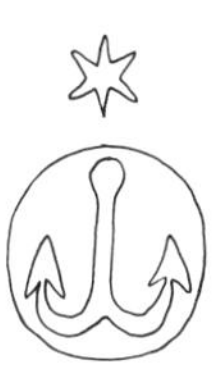
159

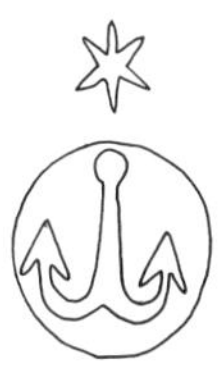
160

161

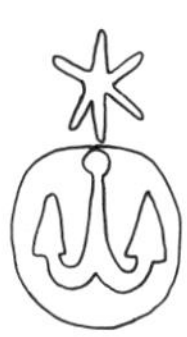
162

163

164

165

166

167

168

169

170

171

172

173

174

175

176

177

178

179

180

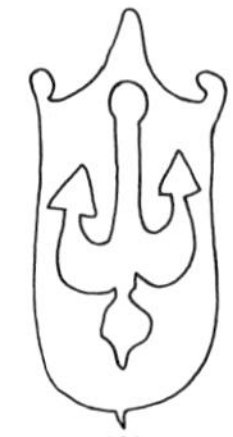
181

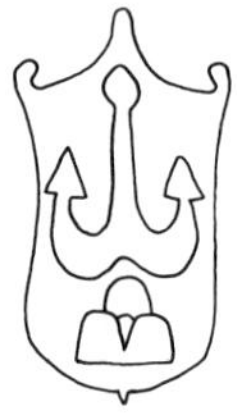
182

183

184

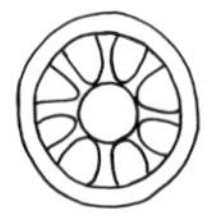
185

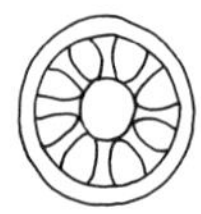
186

187

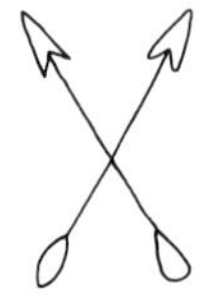
188

189

190

191

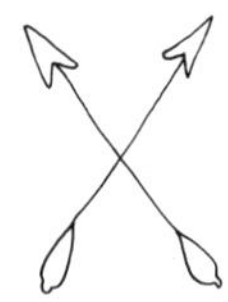
192

193

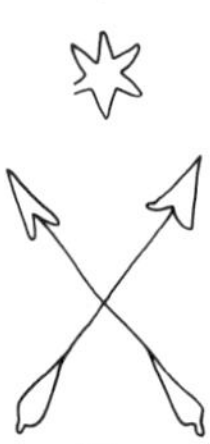
194

195

196

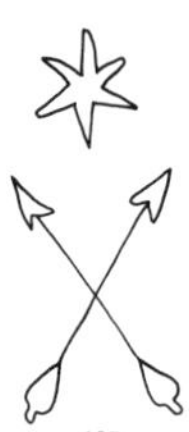
197

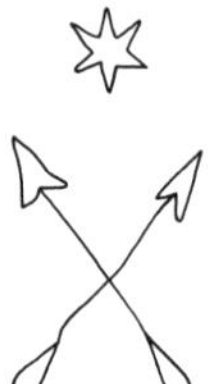
198

199

200

201

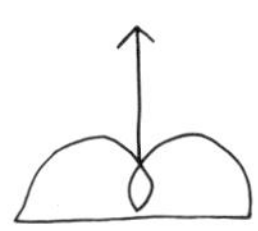
202

203

204

205

206

207

208

209

210

GRAPHIC INDEX

211

212

213

214

215

216

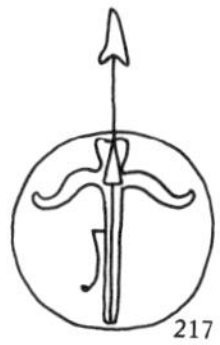
217

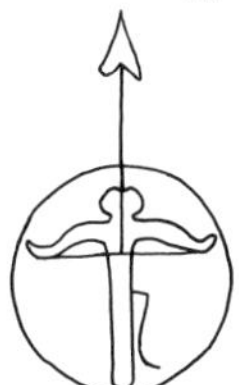
218

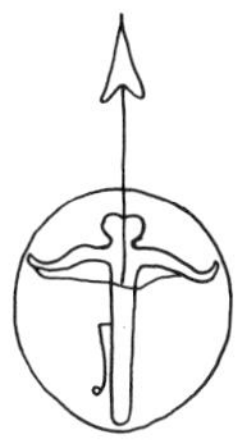
219

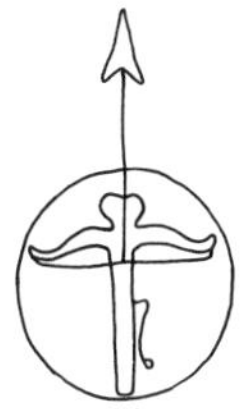
220

221

222

223

224

225

226

227

228

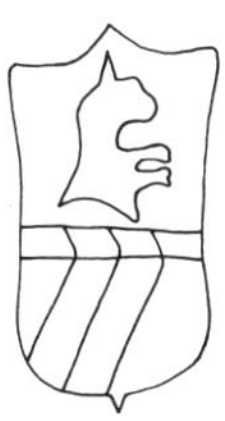
229

230

231

232

233

234

235

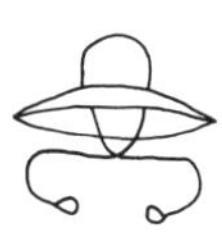
236

237

238

239

240

241

242

243

244

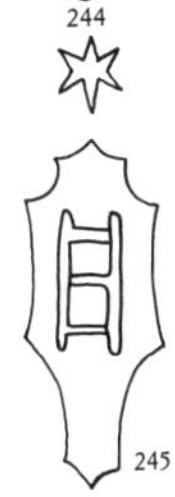
245

246

GRAPHIC INDEX

247

248

249

250

251

252

253

254

255

256

257

258

259

260

261

262

263

264

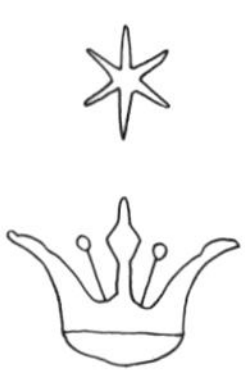
265

266

267

268

269

270

271

272

273

274

275

276

277

278

279

280

281

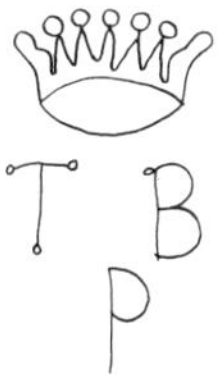
282

283

284

285

286

287

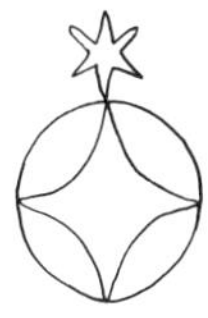
288

289

290

291

292

293

294

295

296

297

298

299

300

301

302

303

304

305

306

307

308

309

310

311

312

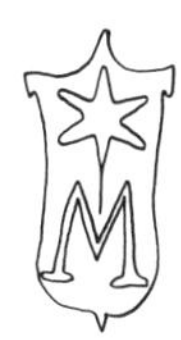
313

314

315

316

317

318

319

320

321

322

323

324

325

326

327

328

329

330

331

332

333

334

335

PRINTED IN FLORENCE
BY “TIPOGRAFIA GIUNTINA”
MAY 1996